Promoting Enterprise-Led Innovation in China

Promoting Enterprise-Led Innovation in China

Chunlin Zhang,
Douglas Zhihua Zeng,
William Peter Mako, and
James Seward

THE WORLD BANK
Washington, DC

1818 H Street NW
Washington DC 20433
Telephone: 202-473-1000
Internet: www.worldbank.org
E-mail: feedback@worldbank.org

1 2 3 4 12 11 10 09

This volume is a product of the staff of the International Bank for Reconstruction and Development / The World Bank. The findings, interpretations, and conclusions expressed in this volume do not necessarily reflect the views of the Executive Directors of The World Bank or the governments they represent.

The World Bank does not guarantee the accuracy of the data included in this work. The boundaries, colors, denominations, and other information shown on any map in this work do not imply any judgement on the part of The World Bank concerning the legal status of any territory or the endorsement or acceptance of such boundaries.

ISBN: 978-0-8213-7753-6
eISBN: 978-0-8213-7939-4
DOI: 10.1596/978-0-8213-7753-6

Library of Congress Cataloging-in-Publication Data

Promoting enterprise-led innovation in China / Chunlin Zhang ... [et al.].
    p. cm.
    Includes bibliographical references and index.
    ISBN 978-0-8213-7753-6 — ISBN 978-0-8213-7939-4 (electronic)
    1. Technological innovations—Economic aspects—China. 2. Technology transfer—China.
I. Zhang, Chunlin, 1957- II. World Bank.
    HC430.T4P76 2009
    338'.0640951—dc22

                                                                                    2009010819

Cover photo: Corbis/National Geographic

Cover design: Naylor Design, Washington, DC

Contents

Figures

Tables

Acknowledgments

This report is a result of a study undertaken by a task team comprising Chunlin Zhang (Task Team Leader), Douglas Zhihua Zeng, William Peter Mako, James Seward, and Wei Zhang. The team was advised by Shahid Yusuf. Substantive input to Chapter 4 was provided by Davin A. Mackenzie. Chapter 2 benefited greatly from background reports prepared by Wenkui Zhang. Background papers were also received from Wei Li and Colin Lixin Xu (innovation performance), as well as Dongsoo Kang (the Korean experience of venture capital industry). Chongqing Productivity Center and Zhejiang College of Industry and Commerce executed the World Bank Chinese SME Innovation Survey in Chongqing municipality and Zhejiang province, respectively, and Shaoqin Zhao processed the survey data. Additional data work was done by Xinxin Kong. Zijing Niu and Li Ouyang provided excellent assistance to the team.

The study has been carried out under the guidance of Director Shen Wenjing of the Ministry of Science and Technology (MOST), and Directors Yang Yingming, Wang Zhongjing of the Ministry of Finance (MOF). The team wishes to express sincere thanks to a large number of government officials, academic researchers, and enterprise managers in Beijing, Hangzhou, and Wenzhou, who have shared their insights with the team in its missions. Among them are Yang Yuecheng, Chen Qian (Torch

Center, MOST); Sun Fuquan, Zhang Ying, Liu Dongmei (China Institute of Science and Technology Development Strategy); Zhen Zijian (Electric Vehicle Special Project Office); Huang Binghua (MOF); Wang Yong (National Development & Reform Commission); Lu Wei (Development Research Center of the State Council); Qian Jinchang (National Bureau of Statistics); Guo Hong, Liu Fudong (Administrative Committee of Zhongguancun Science Park); Wu Zhenyi (Beijing Tsinghua Solar Co., Ltd.); Hong Jiqing, Zhu Zuchao, Zhou Weiqiang, Wang Qun (Science and Technology Department of Zhejiang Province); Chen Weidong (Hangzhou Hi-Tech Investment Guarantee Co., Ltd.); Liu Wei, Chen Jun, Li Xiaopan, Cheng Hong (Wanxiang Group Co.); Jin Wanshu (Hangzhou Kaiyuan Computer Technology Co., Ltd.); Xu Jiasui (Hangzhou ECHO Computer Technology Co., Ltd); Yao Naxin (Hangzhou Focused Photonics, Inc.); Wang Beijiao, Jin Chuanshun (Wenzhou Science & Technology Bureau); Dong Jinxin (Yongjia Pump & Valve Technology Innovation Center); Zhang Xiaozhong (Wenzhou Baoyi Group Co., Ltd.); and Yang Chen (Wenzhou University).

The team is also grateful for insightful comments made by peer reviewers Itzhak Goldberg, Alfred Watkins and Lan Xue, as well as Vikram Nehru, Tunc Tahsin Uyanik, Bert Hofman, Jun Wang, Jae Hoon Yoo, Shen Wenjing, Wang Yuan, Lu Wei, Wei Xiangqun, and Fang Hanting.

The team worked under the general guidance given by Tunc Tahsin Uyanik, Sector Manager, Financial and Private Sector, East Asia and Pacific Region; Homi Kharas and Vikram Nehru, Sector Directors, Poverty Reduction and Economic Management, East Asia and Pacific Region; and David R. Dollar, China Country Director.

Abbreviations

CASS	Chinese Academy of Social Sciences
CPC	Communist Party of China
CSMEI Survey	Chinese Small and Medium-sized Enterprise Innovation Survey
ESTD	early-stage technological development
EVA	economic value added
FDI	foreign direct investment
GDP	gross domestic product
GP	general partner
HEI	higher-education institution
HR	human resources
IPO	initial public offering
ISO	International Standards Organization
IPR	intellectual property rights
IT	information technology
LMEs	large and medium-sized enterprises
LP	limited partner
M&A	merger and acquisition
MII	Ministry of Information Industry
MOF	Ministry of Finance

MOST   Ministry of Science and Technology
MSTQ   measurements, standards, testing, and quality
NBS   National Bureau of Statistics
NDRC   National Development and Reform Commission
NSSF   National Social Security Fund
OECD   Organisation for Economic Co-operation and
                Development
PSU   public service unit
R&D   research and development
RDI   research and development institution
SAC   Standardization Administration China
SASAC   State-Owned Assets Supervision and Administration
                Commission (of the State Council)
SOE   state-owned enterprise
SMEs   small and medium-sized enterprises
S&T   science and technology
TFP   total factor productivity
VAT   value added tax
VC   venture capital

Currency

Y = yuan
$ = U.S. dollar
Exchange rate effective August 8, 2008
$1 = Y 6.8620
Y 1 = $0.1457

Executive Summary

China has made remarkable gains in industrialization and development. Over the past three decades, it has maintained gross domestic product (GDP) growth of about 9 percent per year and lifted more than 400 million people out of poverty. Entering the 21st century, China is determined to ensure the sustainability of its economic and social development, to which the innovativeness of business enterprises is critical. In 2006, the government of China laid out a strategy of enterprise-led indigenous innovation.

Development Challenges

In implementing this strategy, Chinese enterprises must cope with two severe challenges arising from the current stage of development. First, they must derive their competitiveness increasingly from innovativeness. Second, while they are innovating, they must also create jobs to keep the Chinese labor force employed.

Currently, in the international arena, the global competitiveness of China's leading manufacturing sectors, which are often perceived as rising competitors, rests more upon factors other than innovativeness. Most leading Chinese exporters remain manufacturers and assemblers of products without possessing core technologies. In the domestic arena, the sustainability of the growth model that China has followed over the past decades has been called into serious question because of its excessive reliance on capital and resources as opposed to knowledge and innovation. A transformation of economic growth strategy toward one that is more solidly based on efficiency and knowledge is widely recognized as essential to China's long-term prosperity. That transformation has been at the center of the government's "scientific development strategy."

Although such a transformation calls for a greater capacity for indigenous innovation, Chinese enterprises are not fully ready for it. Chinese industry today is a combination of a small number of innovators together with a large number of producers that are engaged in what is called "manufacturing without innovation." A survey covering nearly 300,000 Chinese industrial enterprises of all sizes found that 53 percent of the large enterprises, 86 percent of the medium-sized, and 96 percent of the small did not have continuous research and development (R&D) activities in 2004–06. In 2007, Chinese firms filed 5,470 international patent applications, of which 1,365 were from one company (Huawei). In addition to China's weak capacity for technology creation, it also suffers from a low degree of industrial concentration and a correspondingly large gap in technology and efficiency between leading large producers and small followers.

Basing competitiveness on innovativeness is, however, just part of the challenge facing Chinese enterprises. Full employment is as important to China's continued development as innovation. Chinese enterprises are relied upon for job creation for a labor force of more than 780 million people, of whom more than 80 percent do not have an education background higher than junior secondary school—an educational profile similar to that of the Taiwanese labor force in the 1970s. In other words, mainland Chinese enterprises must be innovative enough to compete with their Taiwanese counterparts in the 21st century international market while creating jobs for a labor force whose educational attainment is approximately at the level attained by the Taiwanese labor force in the 1970s.

Transition Challenges

The severity of the development challenges is compounded by the transitional nature of China's economy and national innovation system. Innovation in pre-reform China, if any, was carried out in a government-led model. Pre-reform state-owned enterprises (SOEs) had neither the motive nor the autonomy to pursue profit maximization, let alone technological innovation. Continuous reform since the mid-1980s has brought China quite far away from the government-led model of innovation, but an enterprise-led and market-based technological innovation system has not yet been established. Creating such a system requires meeting another set of challenges.

First, despite the replacement of government-run R&D institutes by business enterprises as the leading performer of R&D activities in recent years, it is not private enterprises that dominate the R&D landscape now but rather SOEs. Large and medium-sized SOEs accounted for 34 percent of China's R&D expenditure in 2006, compared with 3.5 percent for their domestic private counterparts. This is despite the fact that domestic, private large and medium-sized enterprises filed 1.8 times more patent applications per million yuan of R&D expenditure and owned 1.9 times more patents per 100 scientists and engineers employed than did their SOE counterparts in 2006.

Second, the market system is not functioning well enough to promote innovation. Underdevelopment of market institutions, such as distortions in pricing; weak enforcement of regulations; and barriers to entry, exit, and fair competition tend to discourage firms from investing in innovation. The use of demand-side incentives such as government procurement and standard setting to support innovation is still in an early stage. Operating in such an environment, even profit-maximizing private firms may have only weak incentives to engage in innovation and technological progress, especially when other options are available for short-term profit maximization.

Third, the innovation capacity of the private sector is weak. The last decade has witnessed a spectacular take-off of China's private sector, fueled by a dynamic pace of business creation. From 2000 to 2006, the number of private enterprises increased 1.8 times. Most Chinese private industrial firms are significantly smaller in size than the industry average, are run by inexperienced owners and managers, and operate with relatively low capital intensity and simple technology. Their capacity in technology absorption, adaptation, and creation is often very limited. In 2006–07, the World Bank conducted a survey covering 367 small and medium-sized enterprises (SMEs, mostly privately owned) in two locales—Chongqing and Zhejiang—to assess the constraints on innovation activities that firms of that size may be facing (and also to examine possible differences in that regard between western inland regions, such as Chongqing, and coastal areas, such as Zhejiang). On the overall question of constraints, the survey results indicated that innovative activities of SMEs are frustrated by a shortage of talent, an inability to use external opportunities and resources, difficulties in getting access to quality innovation services, and a lack of capital.

Fourth, the supporting institutional infrastructure, or ecosystem, for a venture capital (VC) industry is not fully developed yet. Despite its relatively early start in the mid-1980s and strong government backing, China's domestic VC industry remains in a nascent stage of development. This is so largely because creating a viable VC industry is more about the creation of an ecosystem than about setting up and capitalizing a number of individual VC firms. And gaps remain in some key dimensions of this ecosystem.

Recommendations

The realization of China's vision to promote enterprise-led innovation will entail concerted actions by government, the corporate sector, and the financial sector. What can the government do? The four basic recommendations of this report are to pursue a balanced strategy, to create the right incentives, to build the capacity of the private sector, and to strengthen the ecosystem for the VC industry.

Pursuing a Balanced Strategy

The success of Chinese enterprises in innovation depends critically on the extent to which the effort serves the overriding objective of sustaining China's economic development. Achieving that orientation entails the adoption of a balanced innovation strategy in a number of dimensions. The first dimension is technological creation versus adaptation and adoption. As in many other developing countries, such as India, China stands to gain from sustained efforts in promoting technology adaptation and adoption. The case for balancing technology creation with adaptation and adoption is particularly strong in view of the pressure for job growth.

The second dimension requires a proper role for the government in promoting innovation as opposed to the market. To ensure that the technologies employed in Chinese industry fully reflect China's comparative advantage, microeconomic decisions on technological innovation are better left to business enterprises and the market.

The third dimension is a balance between R&D expenditure and other factors of innovation such as R&D manpower and a robust infrastructure for science and technology, which are critical to the efficiency of R&D expenditure. As R&D outlays further increase to reach the targeted 2 percent of GDP, a sharper focus on the effectiveness and efficiency of such spending, especially public R&D spending, is highly desirable.

Creating the Right Incentives

In view of the time it takes for market institutions to develop and the private sector to grow, the advisable strategy for the government to create and strengthen incentives for innovation has the following key components:

- Sustained efforts to promote continuous development of the private sector
- Further reform of SOE governance that focuses on board governance
- Further reduction of the scope of state ownership through means such as dividend collection and secondary share offerings
- Greater determination in implementing planned reforms in the areas of pricing energy and natural resources; enforcing laws and regulations on environmental protection; protecting labor rights; pursuing product quality and antimonopoly policies; and removing barriers to entry, exit, and the free transfer of corporate control through merger and acquisition
- Better use of supply-side incentives such as fiscal incentives to encourage pooled R&D efforts locally and globally
- Better use of demand-side instruments such as government procurement and standard setting to raise the demand for innovation, with adequate efforts to guard against the potential risks of rent-seeking and protectionism.

Building the Capacity of Private Enterprises

Although active government involvement in innovation is common in the successful catch-up economies in East Asia, none of the successes can be attributed to a reliance on SOEs. On the contrary, a positive impact from strong government intervention was made possible only by the collective action of private enterprises that were highly motivated to push their technological frontier outward. From that perspective, it is of strategic importance for China to invest in technological capacity building of the emerging private sector, which is now populated with young SMEs run by inexperienced owners and managers operating with relatively low technology.

To move forward, both the private sector itself and the government need to invest more in improving human resources management in private SMEs. The commonly perceived shortage of technological talent is clearly a constraint on private SMEs, but it needs to be interpreted with caution. Analyses suggest that a large part of the problem in hiring and keeping skilled technical personnel is the inadequate internal management of human resources. Private enterprises and the government are advised to consider the following actions:

- Modernizing human resources management, starting with enforcement of labor rights under the Labor Contract Law
- Making use of the legal instruments of confidentiality agreements and competition restrictions to protect technical secrets from being taken by R&D personnel when they leave the firm
- Adapting better to labor market conditions by using relevant services: In particular, local governments could create SME skill development centers to (1) provide SMEs with management and technical training especially related to innovation; (2) provide information on the demand and supply for various skills and the premium on various job categories through close relationships with schools, training institutions, and the labor market; (3) collect and disseminate success stories, especially those from inside China, about the management of skilled employees and the promotion of an innovation culture
- Strengthening policies supporting training and vocational education by reviewing the ceiling on tax-deductible training expenditures (2.5 percent of wage bill) of enterprises and redefining the role of the government in vocational education.

In addition to human resources management, improvements can also be made in facilitating the collaboration of SMEs with knowledge institutions and enhancing innovation services:

- The government could consider initiatives that use more or less SME-specific measures to facilitate SME participation in innovation networks—for example, programs like the innovation brokerage program of the Norwegian TEFT, the innovation voucher program in the Netherlands, and the personnel mobility scheme of the U.K. Business Fellowship.
- The government could also take sectorally tailored actions to promote the development of innovation services, particularly those that are of a public goods nature. Acceleration of the reform of industrial associations is also desirable.

Strengthening the Ecosystem for the VC Industry
Innovation can be financed in multiple ways, of which the supply of external risk capital, that is, capital from outside of firms whose investors

are willing and able to take the risks involved in technology innovation, is a special issue. The VC industry emerged to meet the need for external risk capital. The supply of such capital to innovative firms is a bottleneck of the existing financial system in China. The way forward is for the government to invest more in improving the support system, or ecosystem, for the domestic VC industry. A number of actions can be taken by the government to overcome some key weaknesses of the existing ecosystem:

- Conduct an assessment of the operations of those domestic VC firms created following the newly amended Partnership Law and identify loopholes and weaknesses that require further legislative or policy actions, with close involvement of institutional investors.
- Expand the sources of VC funding by considering policy measures to allow institutional investors to begin investing in domestic VC institutions. Because the risks of VC investing are high, the first step could be to develop a short- and medium-term action plan that would provide a roadmap facilitating investment in private equity and VC funds by institutional investors.
- Build stronger venture partners for investee companies by enhancing corporate governance. The government could organize the formulation of a Code of Conduct for Corporate Governance to facilitate the enforcement of the amended Company Law. It is particularly advisable for the State Council to adopt a regulation to govern the issuance of preference shares.
- Further widen the avenues for exit for venture investments by providing mechanisms for foreign-VC-invested companies to list on both foreign and local exchanges; and further improve the domestic listing process by, for example, further reducing the time required for application, introducing greater transparency, and reducing government management of listing volumes.

A cross-cutting area of the VC industry is the role of the government. As is the case in many other countries, there is very limited transparency to the government's direct interventions in China's VC industry and little rigorous empirical study of the impact of those interventions. Nonetheless, the practical experiences of other countries with government intervention are mixed. China could consider giving priority to the strengthening of the VC ecosystem as mentioned here while also assessing the merits of direct involvement in the VC industry and the most appropriate role for government intervention in innovation financing more generally.

This report has also touched upon a number of subjects that call for further study. Those issues are key to a better understanding of the challenges and viable policy options facing China in its effort to develop into an "innovative nation." Examples of subjects for further study include the evaluation methodology for public R&D spending, the route to job-creating innovation, ways to strengthen innovation management within firms, the development of innovation services, and the role of the government in supplying external risk capital.

CHAPTER 1

Pursuing a Balanced Strategy

China's economic performance has been spectacular for the three decades since it began reforming and opening up in the late 1970s. Over that period it has maintained gross domestic product (GDP) growth of about 9 percent a year and lifted more than 400 million people out of poverty. Can this record of growth and poverty reduction be sustained? This is the central challenge facing China today. And the innovativeness of its business enterprises is critical to China's meeting that challenge.

Indeed, building a broad-spectrum system of innovation is widely viewed as the central plank of growth strategies for industrialized and industrializing countries alike (Yusuf, Wang, and Nabeshima 2009). The government of China is fully aware of the critical importance of innovation to economic performance. In early 2006, after several years of intensive consultation and research, the government announced its *Outlined National Program for Medium and Long Term Development of Science and Technology (2006–2020)* (the 2006 S&T program). The ruling Communist Party of China (CPC) adopted a "Decision" implementing the 2006 S&T program.

The 2006 S&T Program

The 2006 S&T program established an innovation strategy for the next 15 years consisting of four pillars: (1) "indigenous innovation" (increasing domestic innovation capacity), (2) a "leap-forward in key areas" (concentrating resources to achieve breakthrough in priority areas), (3) "sustaining development" (meeting the most urgent demands of economic and social development), and (4) "setting the stage for the future" (getting prepared for future development with a long-term vision).

The first pillar, indigenous innovation, codifies the determination to reduce China's dependence on foreign technology and is the central theme of the new strategy. Indeed, the 2006 S&T program has put forth a measure of dependence on foreign technology called the *dependence ratio*, defined as the expenditure on technology imports as a percentage of the sum of research and development (R&D) expenditure and technology import expenditure (Wu and Gao 2007). The program set a goal of reducing the dependence ratio to 30 percent by 2020.

The notion of indigenous innovation is complemented in the 2006 S&T program by a greater emphasis on the role of business enterprises in technological innovation. Both the 2006 S&T program and the CPC's Decision called for a "leading role" (*zhuti*, or central role) for enterprises in technological innovation. In stating the guiding principles of S&T system reform, the 2006 S&T program identifies "a technological innovation system led by enterprises" as the "point of breakthrough." The CPC's Decision elaborates on how indigenous innovation is to be achieved:

> The key to increasing indigenous innovation capacity is to strengthen the leading role of enterprises in technological innovation to build up a technological innovation system that is led by enterprises, guided by the market, and characterized by collaboration of industries, HEIs and research institutes. More effective measures must be taken to create a more conducive environment that enables enterprises to play a leading role in R&D expenditure, technological innovation activities as well as the application of results of innovation.

Purpose and Scope of This Study

This study is designed to assist the government of China in implementing its strategy of enterprise-led technological innovation; the ultimate goal of that strategy, and of this study, is to improve the sustainability of China's economic development and poverty reduction. The rest of this chapter sets the stage for the report: it reviews the innovation achievements and challenges of Chinese enterprises and discusses several strategic issues. Chapter 2 concentrates on incentives for Chinese enterprises to innovate by examining potential ways to strengthen corporate governance of state-owned enterprises (SOEs), external market incentives, and the demand for innovation. Chapter 3 turns to building the capacity of the emerging private sector to create and absorb technology; it focuses on issues regarding human resources management and collaboration with knowledge institutions and innovation services. Chapter 4 addresses the financing of innovation and the development of an ecosystem conducive to the domestic venture capital industry. Chapter 5 concludes by summarizing the recommendations for government action that were presented in each chapter.

A Broad Definition of Innovation

Innovation can be defined in many ways. It can be narrowly defined as the creation of technology that is new to the world. It can also be defined broadly, "to encompass the processes by which firms master and get into practice product designs and manufacturing processes that are new to them, if not to the universe or even to the nation" (Nelson and Rosenberg 1993). In certain cases, adoption of a technology that is only "new to the firm" could appear too far from the notion of "innovation." However, on balance, when performance in economic development is the central concern, it is the broadly defined concept of innovation that matters more (Nelson and Rosenberg 1993).

This study adopts the broad definition to cover two distinct sets of innovative activity by firms. The first set is technology *creation* from a global perspective, that is, the design and production of technologies with worldwide significance. Technology creation is often conducted by large corporations and small, creative firms that are approaching technological frontiers in a global context. In contrast, the second set is more of the nature of *adoption* or *adaptation*. Such activity can be local improvement based on the adoption of technologies that are more or less available worldwide or locally, or it can be the building-up of competitive activities with some adaptation made to existing technologies.[1] Adoption and adaptation can be conducted by any enterprise, including those that are relatively far from global technological frontiers and not ready to engage in technology creation.

The Scale of Chinese Innovation

China has dramatically scaled up its investment in R&D over the past ten years. From 1995 to 2006, the full-time equivalent of R&D personnel was doubled, from 0.75 million to 1.5 million person-years (NBS and MOST 2007, p. 4). R&D expenditure as a share of GDP (R&D intensity) rose from 0.5 percent to 1.42 percent. With fast GDP growth (about 9 percent per year), total R&D expenditure increased sharply, 5.5 times in real terms (figure 1.1). The high annual rate of growth of R&D spending is a clear manifestation of a strong catch-up movement relative to the United States, Japan, the members of the European Union (EU), and the Russian Federation (figure 1.2). A closer look at the data suggests that the strong increase in R&D expenditure since 2001 (for which more detailed data are available) was mainly driven by large and medium-sized industrial enterprises (LMEs) (table 1.1).[2]

Data recently released by the NBS and NDRC on S&T activities of LMEs (available only for 2005 and 2006) permit a look into the structure of the R&D expenditure of LMEs in more detail. According to the data for 2006 (table 1.2),

- More than 80 percent of the total R&D expenditure of industrial LMEs is concentrated in 10 sectors, of which the top 4 account for more than 50 percentage points: electronics, transportation equipment, electrical machinery, and iron and steel.
- Domestic LMEs contributed 72.7 percent of the total R&D spending of all LMEs, and large enterprises contributed 66.6 percent. Overall,

Figure 1.1   China's R&D Expenditure, 1995–2006

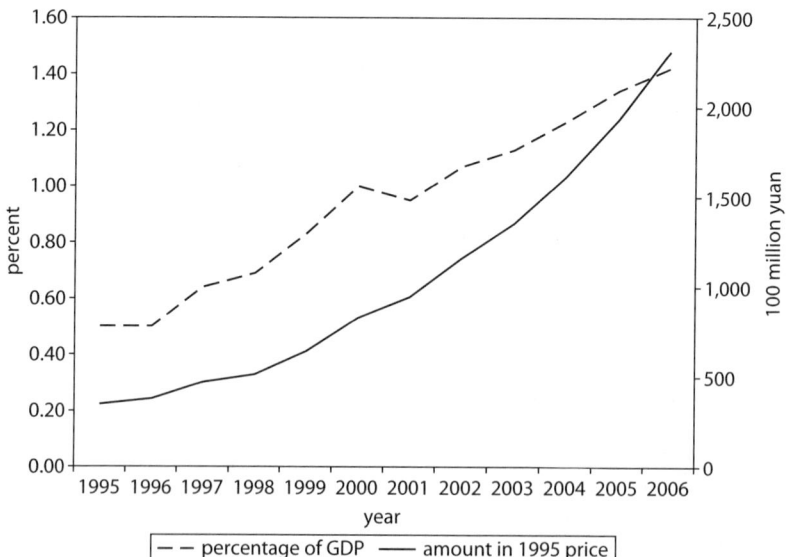

Figure 1.2   R&D Intensity in 2004 and Annual Average Growth Rate of R&D Intensity, 1999–2004

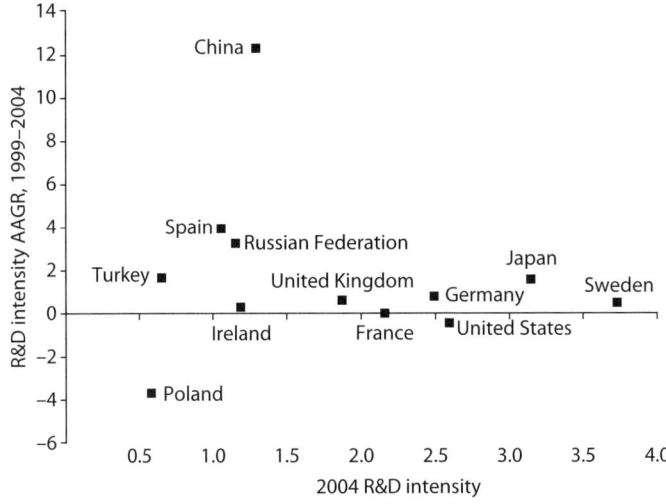

Source: Serger and Breidne 2007, p. 139.
Note: Intensity is expenditure as a percentage of GDP. AAGR = average annual growth rate.

Table 1.1   Annual Increase of China's R&D Expenditure, by Performing Sector, 2001–06

(percent except as noted)

Year	Total increase (Y billions)	Industrial enterprises			RDIs	HEIs	Other
		All	LMEs	SEs			
2001	14.68	63.3	60.5	2.8	20.8	17.5	−1.6
2002	24.52	64.4	48.8	15.5	25.6	11.5	−1.4
2003	25.20	68.4	63.0	5.4	18.9	12.6	0.0
2004	42.67	82.9	54.7	28.2	7.7	9.0	0.4
2005	48.36	74.4	61.2	13.2	16.8	8.6	0.2
2006	55.31	83.3	68.7	14.6	9.8	6.2	0.7

Source: NBS and MOST 2007, p. 7.
Note: LMEs = large and medium-sized industrial enterprises; SEs = small industrial enterprises; RDIs = Research and Development Institutes; HEIs = higher-education institutions; Y = yuan. For definition of firm sizes, see text note 2.

the role of foreign-direct-investment (FDI) enterprises and medium-sized enterprises are of secondary importance.
• Sector by sector, the role of FDI enterprises varies significantly. For example, they account for 51.6 percent of total R&D expenditure in the electronics sector and 35.6 percent in transportation equipment manufacturing sector.

Table 1.2   Structure of R&D Expenditures of Chinese LMEs, Large and Domestic, by Selected Sectors, 2006

(percent)

Sector	All LMEs	Share in total of all sectors		Share in total of each sector	
		Domestic	Large	Domestic	Large
All sectors	100	72.7	66.6	n.a.	n.a.
Communication equipment, computers, and other electronics	21.4	10.3	14.9	48.4	69.9
Transportation equipment	13.7	8.8	10.4	64.4	75.5
Electrical machinery	10.2	7.7	6.0	75.0	58.4
Black metallurgy	9.9	9.5	9.4	95.0	94.2
General equipment	6.3	4.5	3.2	70.3	51.1
Chemical	6.0	5.4	3.3	90.5	55.7
Special equipment	4.7	3.9	2.4	84.2	50.6
Nonferrous metallurgy	3.4	3.1	2.6	92.1	76.5
Pharmaceutical	3.2	2.4	1.2	73.4	36.3
Coal mining	2.3	2.3	2.2	100.0	97.6
Ten sectors, total	81.2	57.9	55.6	n.a.	n.a.

Source: NBS and NDRC 2007.

Note: n.a. = not applicable.

- The R&D spending of medium-sized enterprises is mostly significant in the pharmaceutical sector, accounting for 63.7 percent. High shares are also found in special equipment manufacturing (49.4 percent), general equipment manufacturing (48.9 percent), and the chemical industry (44.3 percent).

Chinese enterprises invest in broadly defined innovation in a number of ways. Besides R&D spending, the most notable way is through technology import. China's industrialization since 1949 has relied heavily on the import of foreign technology. Starting in the late 1970s, China opened up to the rest of the world and pursued an active strategy of technology import through both trade and foreign direct investment. During 1978–2002, China spent $225.7 billion on the import of technology.[3] In 2006 alone, the total import of technology amounted to $22.02 billion, or 57 percent of total R&D spending; of that amount, $8.68 billion was paid for technology licenses and $4.29 billion was related to FDI in joint ventures (NBS and MOST 2007, p. 298). Over the past decade, however, Chinese enterprises have been shifting their spending from technology import to R&D (figure 1.3).

Figure 1.3   Expenditures on R&D, Technology Import, and Technology Absorption by Chinese LMEs, 1995–2006
(*percentage of sales revenue*)

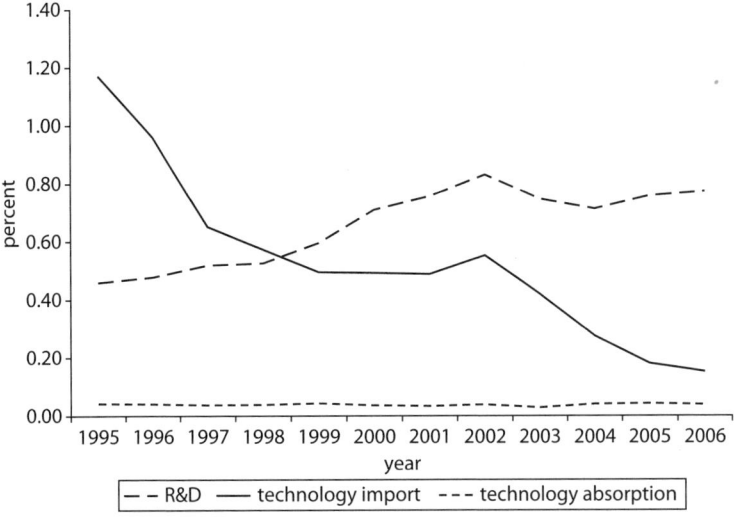

Source: NBS and MOST 2007, p. 92.

In a complement to technology import, Chinese enterprises also spend on the absorption of imported technology, part of which spending is counted as R&D in official statistics. Investment in the absorption of imported technologies by Chinese LMEs has been insignificant (0.03–0.04 percent of sales revenue) and stagnant over the past decade (figure 1.3).

The Achievements of Chinese Innovation

With increasingly intensive inputs of R&D and other resources, Chinese enterprises have recorded remarkable achievements in the transfer, absorption, and generation of technology, among them a large increase in manufacturing and in its technological sophistication, a rise in the ability to innovate at home (and for a few firms, at the international state of the art), and a rise in the knowledge intensity of the overall economy.

The Expansion of Manufacturing and Technological Capacity

The continuous import and absorption of foreign technology have enabled Chinese enterprises to dramatically expand their manufacturing capacity. The motor vehicle industry serves as a case in point. In 1980,

China produced only 0.22 million motor vehicles and only 5,400 cars, all of which were the same models introduced from the former Soviet Union in the early 1950s, when the First Automobile Works was established. By 2006, motor vehicle production had jumped to 7.3 million units, including 3.9 million cars (NBS 2007a, p. 555). China's share in global output of motor vehicle production increased from 0.4 percent in 1978 to 12.2 percent in 2007 (DIE/DRC and others 2008, p. 290). An assessment by the Chinese Academy of Social Sciences (CASS) found that at the end of the 20th century, key products of the Chinese motor vehicle industry had reached the 1980s level of international technology, and the technological distance of Chinese motor vehicle products from the international frontier was reduced from 20–30 years to 10–15 years (CASS 2004, pp. 220–21). The household electric appliances industry is another example of dramatic technological progress. The history of this sector "is a history of technology import" (CASS 2004, p. 62). Technology absorption enabled the industry to rise from scratch in about 10 years. Now Chinese producers of household electric appliances (including television sets, refrigerators, washing machines, and air conditioners) account for one-fourth of world production.

China's expansion of its manufacturing capacity has been aided by two closely related trends. First, because of the maturing of certain technologies and the parallel growth of consumer markets, many manufactured products have become standardized commodities. Second, the very process of "commodification" has been supported by the codifying of the associated technologies, some embedded in equipment, others available from suppliers. These changes have made it easier to absorb new production methods and quickly achieve a high level of efficiency (Yusuf and Nabeshima 2007).

Chinese enterprises have mastered a range of relatively advanced technologies and thus have improved their technological capacity. For example, in the steel industry, key aspects of the production technology employed by major producers such as Angang, Baogang, Baosteel, and Wugang have reached or come close to the international frontier. Almost every large or medium-sized steel producer has one or two production lines equipped with internationally advanced technology. The overall level of technology adopted by the Chinese steel industry is now much higher than it was two decades ago (CASS 2004, p. 63). In the machinery industry, computer-aided design (CAD) has been popularized among machinery manufacturers. Through the absorption of imported technology, Chinese producers have been able to develop and manufacture critical machinery for a wide range of industries, including heavy equipment for open-pit coal mining sites, power transmission and transformation equipment at the scale of 500 kilowatts, and large-scale continuous casting and cold rolling equipment for the steel industry (CASS 2004, pp. 58–59).

A Greater Ability to Innovate at Home

Some Chinese enterprises are now becoming increasingly innovative. That progress is partly reflected by the sharp increase in the number of patents granted to domestically funded Chinese enterprises in the 1995–2006 period (table 1.3).[4] Although the 13-times rise in total number of patents granted during the 10 years is impressive, the even sharper increase in the number of patents for invention, from 205 in 1995 to 9,433 in 2006, is particularly informative of the efforts and pace of Chinese enterprises in moving up the ladder from technology absorption to creation. The share of "new products" (as defined by official statistics) in total sales revenue is another indicator of the innovativeness of enterprises that is widely used in official Chinese statistics; LMEs have managed to raise that share from 8.5 percent in 1995 to 14.8 percent in 2006 (NBS and MOST 2007, pp. 7 and 92–93).

A small number of Chinese enterprises have reached or are approaching the international technological frontier with their growing ability to create technology. That is particularly the case in the electronics industry, where Chinese firms have reached the international frontier in 3G (third generation) technology. Leading firms such as Huawei and ZTE have become key international players (CASS 2004, p. 251). In 2007, Huawei moved up nine places to become the fourth-largest patent applicant under the Patent Cooperation Treaty (PCT), with 1,365 applications published.[5]

Table 1.3   Patents Granted by Chinese Authorities to Domestically Funded Chinese Enterprises, by Type of Patent, 1995–2006

Year	Total	Invention	Utility models	Design
1995	5,386	205	2,627	2,554
2000	31,319	1,016	12,821	17,482
2006	76,379	9,433	35,667	31,279
Increase as a factor of the initial level, 1995–2006	13.2	45.0	12.6	11.2

Source: NBS and MOST 2007, p. 823.

A Rise in Knowledge Intensity

The overall structure of the Chinese economy has become increasingly knowledge based, as reflected by the share of the relatively knowledge-intensive industrial sectors in the economy. Chinese official statistics define five sectors as "high-tech industries":

- pharmaceuticals
- aircraft and spacecraft
- electronics and communication equipment
- computers and office equipment
- medical equipment and measuring instruments.

The rising share of output of those five sectors in GDP suggests a strong rising trend of knowledge intensiveness (figure 1.4).

The Contribution of Innovation to China's Current Economic Success

China's unprecedented performance in economic growth and poverty reduction has undoubtedly been sustained by the technological progress of its business enterprises. A number of studies have used growth accounting

Figure 1.4   Value Added of "High-Tech Industries" as a Percentage of China's GDP, 1995–2006

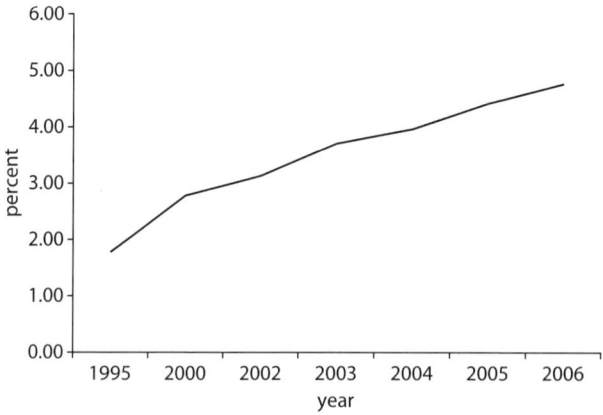

Sources: NBS 2007a, p. 57; NBS, NDRC, and MOST 2007, p. 3.

Note: In China's official statistics, the high-tech industries are (1) pharmaceuticals, (2) aircraft and spacecraft, (3) electronics and communication equipment, (4) computers and office equipment, and (5) medical equipment and measuring instruments.

to identify the sources of China's rapid development since the late 1970s. Although findings differ because of variations in assumptions, they have converged on the significant contribution of the growth of total factor productivity (TFP). Estimates of TFP growth during the reform period range between 2 percent and 4 percent per year. On the basis of revised GDP data, a recent study found that China's TFP growth was 3.8 percent during 1978–93 and 3 percent during 1993–2005 (Kuijs 2006, pp. 3–4). Such rates are very high by international standards. For example, the TFP growth rate was 2 percent in Japan in 1960–89; 1.7 percent in the Republic of Korea in 1966–90; 2.1 percent in Taiwan, China, in 1966–90; and 0.4 percent in the United States in 1960–89 (HKMA 2006).

China's high rate of TFP growth is, of course, the result of many factors, particularly institutional changes brought about by reform and opening up. However, technological progress is certainly one of the key sources. In a background paper prepared for this study, Li and Xu (2007) drew on a survey conducted jointly by the World Bank and the National Bureau of Statistics (NBS) that covered 12,400 enterprises in 120 Chinese cities during 2002–04. Li and Xu used the survey data to measure the private return to R&D spending of Chinese enterprises. They found that the marginal return for individual (firm) R&D investment in the period was substantial, ranging from 0.69 to 0.74; investing in a dollar of R&D yielded an increase in TFP of about $0.70. That finding is robust with respect to alternative controls for firm, industry, and city characteristics. It is also robust to the endogeneity of R&D intensity. The return to R&D investment of 0.70 is larger than what was found at the firm level between 1973 and 1980 for the United States (0.27 to 0.41) and for Japan in the same period (0.30 to 0.56) with similar specifications (Griliches 1998, p. 203). The return is comparable with that in Japan in 1976–77 and 1977–78, when the marginal return was 0.85 and 1.01 respectively; and with the United States in 1978–79, when the number was 1.28. The study also found strong spillover effects.

The Innovation Challenges Faced by China

Innovation is a race without an end. Even the most advanced nations have their own challenges and must keep innovating to stay in the race. China's past achievements are undoubtedly impressive. However, they do not change the fact that China is still a late-comer compared with technologically advanced nations, and the gap between it and the international technological frontier remains substantial.

An Epitome of Chinese Industry

The innovation challenges China is facing can be illustrated by the case of Logitech International SA. According to a 2004 *Wall Street Journal* article, Logitech—a Swiss-American company headquartered in California—shipped 20 million computer mice to the United States in 2004 (Higgins in Wu 2007). They were assembled in Suzhou, a Chinese city in prosperous Jiangsu province, which surrounds Shanghai.

> One of Logitech's big sellers is a wireless mouse called Wanda, which sells to American consumers for around $40. Of this, Logitech takes about $8, while distributors and retailers take $15. A further $14 goes to suppliers that provide Wanda's parts: A Motorola Inc. plant in Malaysia makes the mouse's chips, and America's Agilent Technologies Inc. supplies the optical sensor. Even the solder comes from a U.S. company, Cookson Electronics, which has a factory in China's Yunnan province next to Vietnam.
>
> Marketing is led from Fremont, California, where a staff of 450 earns far more than 4000 Chinese employed in Suzhou. China's take from each mouse comes to a meager $3, which covers wages, power, transport and other overhead costs. . . .
>
> Logitech, like most tech, toy and textile companies with plants in China, employs mostly young women such as Wang Yan, an 18-year-old from the impoverished rural province of Anhui. She is paid $75 a month to sit all day at a conveyor belt plugging three tiny bits of metal into circuit boards. She does this 2,000 times a day. . . .
>
> This is her second stint in a factory. Before coming to Suzhou, she skipped school to become an underage worker at an electronics plant . . . She complains about her salary but isn't going back to her village. That would mean only "eating bitterness," she says.

To a significant extent, the Logitech operation is the epitome of Chinese industry in the 21st century: a quickly rising competitor in the global market that derives its strength from sources other than innovativeness. All technological late-comers started their catch-up by importing, adopting, and absorbing foreign technology. The successful ones were able to ensure that by the time they were perceived as threats to incumbents in the global market (and, naturally, their terms of technology importation had become less favorable), their domestic firms had developed a capacity for indigenous innovation sufficient to sustain their movement toward the international technological frontier. That "crossover" point is largely where Japan found itself in 1960s (Odagiri and Goto 1993, p. 87). China now seems to be rapidly approaching that critical point. However, the capacity of Chinese enterprises for indigenous innovation appears inadequate to cope with the challenge.

Indeed, competition in the global market is just part of the story. In the domestic arena, the sustainability of the growth model that China has followed over the past three decades has been called into serious question for its excessive reliance on inputs of capital and natural resources as opposed to knowledge and innovation. A transformation of its economic growth strategy into one that is founded more on efficiency and knowledge is widely recognized as essential to China's long-term prosperity (for example, Wu 2007) and has been at the center of the Chinese government's development strategy known as the "scientific development strategy." The challenge, in essence, is that if China's competitiveness is to be sustained, Chinese enterprises must not only be competitive—they must also become more innovative.

Deriving Competiveness from Innovativeness

It is true that some Chinese enterprises are increasingly innovative, that a small number of them are even approaching the international technological frontier, and that the general level of technological progress of Chinese industry has been rising over the past three decades. Nonetheless, the overall picture remains that the global competitiveness of China's leading manufacturing sectors rests upon low input costs, scale of production, technology absorption, speed of response to market demands and customer orders, and increasing attention to the quality of products (Yusuf, Wang, and Nabeshima 2009).

Most leading Chinese enterprises remain manufacturers and assemblers of products without possessing core technologies. Even in joint ventures, core technologies mostly remain controlled by the foreign partners. China's export growth has been largely based on the expansion of low-wage manufacturing using imported components, equipment, and technology (OECD 2007, pp. 12–15). As found by a recent study (Amiti and Freund 2008), China's exports increased more than five times during 1992–2005, and its structure transformed dramatically. The shares of agriculture and the manufacture of soft goods (such as textiles and apparel) have declined significantly, and the share of hard-goods manufactures (such as consumer electronics, appliances, and computers) has grown. On the surface, China appears to be changing its comparative advantage. However, a large component of its export growth in hard manufactures has been growth in processing trade. The skill intensity of China's exports remains unchanged since 1992 once one accounts for processing trade.

One might find it hard to reconcile the lack of innovation in the Chinese industry with the stories of rapidly rising Chinese innovators such as Huawei. Indeed, Chinese industry today is a combination of a small number of innovators together with a large number of producers who are engaged in "manufacturing without innovation." In an informal survey covering 299,995 Chinese industrial enterprises for the period 2004–06, the NBS found that 53 percent of the large enterprises, 86 percent of the medium-sized ones, and 96 percent of small ones did not engage in continuous R&D activities.

As a result, the overall capacity for technology creation by domestically funded Chinese enterprises is weak despite the presence of some catch-up momentum. The problem can be seen in part in the relatively low share of invention in total patents granted by the Chinese State Intellectual Property Office in comparison with foreign firms. In 2006, invention accounted for 11 percent of patents granted to domestic enterprises and 74 percent to foreign enterprises (OECD 2007, p. 32). Internationally, Chinese firms filed 5,470 international applications under the Patent Corporation Treaty (PCT) in 2007, up 38.5 percent from 2006 and 3.2 times from 2003, which is nothing short of remarkable. Nonetheless, China's 2007 total represents only 3.5 percent of the international total, compared with 33.6 percent for the United States, 17.5 percent for Japan, and 4.5 percent for Korea. If those from Huawei are excluded, the number of Chinese applications drops to 4,105, and China's share in the international total drops to 2.6 percent. Indeed, the non-Huawei total of Chinese applications is about the same as the sum of just two leading non-Chinese companies, Matsushita Electric of Japan (2,100 applications) and Koninklijke Philips Electronics of the Netherlands (2,041 applications) (WIPO 2007).

In addition to their generally weak capacity for technology creation, enterprises in China also have large gaps in their capacity for technology adoption and adaptation. The gaps are partly demonstrated by the technological disparity between large and small firms. Because of a host of factors such as their young age and limited access to finance, small enterprises tend to equip themselves with simpler technology and be less capital intensive. And Chinese industries are generally characterized by a low degree of concentration and a large gap in technology and efficiency between leading large producers and small followers. An example is the Chinese cement industry, which is now the world's largest producer (box 1.1).

Box 1.1

China's Cement Industry

China is the world's largest cement producer, accounting for more than 48 percent of world cement production in 2005. Despite its dominance, China's cement industry remains fragmented (5,000 producers in 2005 with an annual revenue exceeding Y 5 million) and inefficient. The average plant size in 2005 was about 220,000 tons per year, far below the international standard. Despite continued government efforts to modernize the industry, only 45 percent of total output in 2005 was produced with new suspension preheater (NSP) technology, a now-mature technology developed in the 1970s. The rest was produced with vertical shaft kilns and other less-efficient technologies, which consume 39–111 percent more energy than NSP. In 2006–07, at least 80 million tons of technologically back-ward capacity was phased out,[a] and the share of NSP-based cement in total cement production rose to 55 percent.[b]

However, the potential for further improvement remains high. Leading Chinese producers such as Anhui Conch have almost completely abandoned non-NSP tech-nologies. As of year-end 2004, Conch had 12 NSP clinker production bases in east-ern and southern China; those plants had a total clinker capacity of 37.9 million tons per year, 95 percent of Conch's overall clinker capacity. Conch introduced the first 8,000-tons-per-day NSP kiln in China in 2003 and the first 10,000-tons-per-day NSP kiln in China in 2004. At that time, only 7 clinker production lines in the world had a capacity of 10,000 tons per day (the largest capacity then technologically feasible), and of those, Conch owned 3. In 1998, Conch became the first cement producer in China to commission a heat recovery power plant on a cement kiln. The recovery plant has since produced half of the kiln's power consumption internally, and the use of power from the national grid was correspondingly reduced.

Sources: IFC 2006; NDRC 2006.
a. "Achievements in 2007 and Priorities for 2008 of the Construction Materials Industry." http://www.c-bm.com/news/2008/1-10/B9740705.htm.
b. News articles at www.sz-sinoma.cn/UploadPDF/UploadPDF/200737132754291.pdf; info.bm.hc360.com/2007/04/02142450227.shtml; and www.bm.cei.gov.cn/allfile/12/2008021815085812023.asp.

Creating Jobs while Innovating

Chinese enterprises face an additional challenge, which is also highlighted by the Logitech story: the need for job creation for unskilled laborers such as the 18-year-old girl, Wang Yan. In other words, Chinese enterprises must not only become fast enough to catch up with the international frontier in technology and knowledge intensity to remain competitive,

Figure 1.5   Educational Attainment of China's Labor Force: A Comparison of the Mainland with Taiwan, China, 1978 and 2006

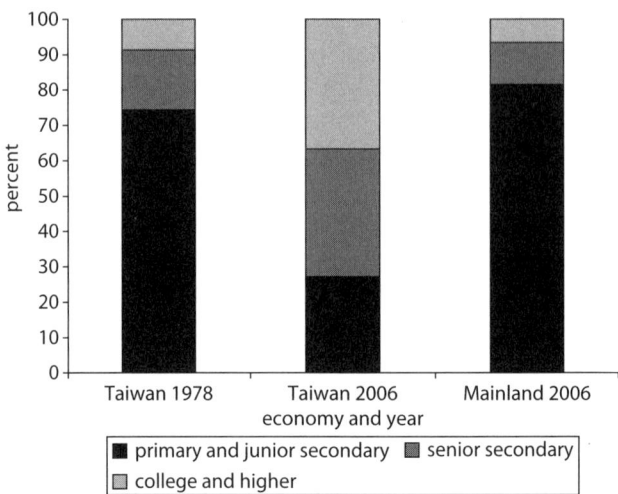

Source: For mainland China, NBS 2007b, table 1-44; for Taiwan, China, http://www.stat.gov.tw/ct.asp?xItem= 17144&ctNode=517.

but must also make sure the nation's huge labor force is properly employed. They must be job creating as well as innovative.

The challenge of achieving job-creating innovation is particularly formidable. As of 2006, China had a labor force of 782 million, of which 82 percent did not have an education higher than junior secondary school. The severity of the challenge is most stark in comparison with Taiwan, China (figure 1.5). The data imply that mainland Chinese firms as of 2006 had to be innovative enough to compete in the international market with Taiwanese firms while creating jobs for a labor force whose educational attainment was similar to that of Taiwan's in 1978.

Pursuing a Balanced Strategy

The government of China has laid out its strategy of enterprise-led indigenous innovation in its 2006 S&T program. In view of the nature of the challenges facing Chinese enterprises, a number of issues appear to be crucial to the successful implementation of the government's strategy.

The overarching issue is the clarity of the objective of innovation. As important as it is, innovation is only a means to an end. The success of

Chinese enterprises in promoting innovation depends critically on the extent to which efforts are oriented to the general objective of sustaining China's economic development. The temptation of some alternatives, such as being first, being high-tech, and being indigenous, is often strong and needs to be resisted. Once the sustainability of economic development is set as the objective, several issues arise from the reality of Chinese enterprises.

Creation vs. Adaptation and Adoption

The first issue is the balance between technology creation, on the one hand, and adaptation and adoption, on the other. This distinction has important implications for Chinese enterprises, because the challenges facing them are dual in nature. Some Chinese enterprises are already facing the challenge of technology creation. That challenge has intensified as Chinese enterprises have become more and more exposed to international competition; a small number of leading enterprises have reached or have come very close to the global technological frontier and can no longer rely on technology transfer to maintain their competitiveness. However, in China as well as in many other developing economies, the majority of enterprises is far from the global, or even national, technological frontier and engages in innovation only to catch up (Gill and Kharas 2007, p. 98).

The dualism in the Chinese economy is perhaps not as stark as in India (box 1.2). Nevertheless, it exists in China, not only in the industrial sector, as shown in the previous section, but also in the economy as a whole. As in the case of India, China stands to gain from a broad interpretation of innovation and sustained efforts in promoting technology adaptation and adoption. Although that kind of innovation may not appear to be as important as technology creation and therefore may be less likely to be given high priority in practice, adaptation and adoption are nonetheless as critical as creation in terms of the impact on the sustainability of economic development. Also, the degree of success to be had from adaptation and adoption can be as spectacular as that from creation, as shown by the case of the Wangxiang Group (box 1.3).

The enormous challenge of job creation in China makes the achievement of a balance between technology creation and technology adaptation and adoption even more essential. Obviously, technologies that maximize job creation are not necessarily those that are the most advanced. Rather, they are those that maximize the profit of the relevant firm under current market conditions, which include, among other things,

Box 1.2

India Stands to Gain More from Absorption than Creation

A recent World Bank study found a stark dualism in the Indian economy. At one extreme, India is a nuclear and space power and is increasingly becoming a top global innovator in certain key economic sectors, including biotechnology, pharmaceuticals, automotive components, information technology, software, and information technology–enabled services. At the opposite extreme, India largely remains a subsistence economy, with 60 percent of its workforce engaged in agriculture. Less than 3 percent of the workforce is employed in the formal private sector. In the 2004–06 period, real GDP has grown more than 8 percent a year. Growth has been driven by a jump in export-oriented, skill-intensive manufacturing (pharmaceuticals, petrochemicals, and auto parts and assembly) and services (information technology, business services, and finance) and has been accompanied by a jump in innovation activities. However, a tremendous dispersion in productivity levels remains, both within and across economic sectors. Roughly 90 percent of the workforce is underemployed in low-skill, low-productivity, low-income activities.

India's dual economic structure and wide dispersion in productivity levels call for a broader interpretation of innovation. India has more to gain from economy-wide productivity increases from the diffusion and absorption of existing knowledge than from the creation and commercialization of new knowledge. If all enterprises in India could costlessly reach the level of national best practice for the knowledge already used in India, the output of the economy could increase more than fivefold.

Source: Dutz 2007 (overview and chapter 1).

characteristics of the available labor supply in the market. The implication of some labor characteristics, such as educational attainment, for the technology choices of Chinese enterprises is a subject that requires further research. However, it seems clear that when they are driven by the profit motive, most Chinese enterprises are likely to stay with labor-intensive technologies for some years to come, a tendency that highlights the importance of innovation in the form of adaptation and adoption.

How can Chinese enterprises make themselves competitive, innovative, and job creating all at once? That combination necessarily involves a complex set of equations whose solution lies in a set of technology choices by enterprises. For example, in Taiwan, China, in the 1950s–70s, the solution to that set of equations turned out to be a transition from

Box 1.3

Wangxiang: Incremental Innovation in the Service of Long-Term Goals

In 1979, when Wangxiang was founded in a small village in Zhejiang province, it was a township enterprise producing and repairing metal tools with the simplest traditional technology. By 2006, the Wangxiang Group reported revenues of more than Y 30 billion, and it provided high-quality jobs to more than 24,000 employees. In 1984, Wangxiang became the first Chinese auto company to export to the United States, and 10 years later, it became the first Chinese auto firm to establish an American subsidiary. In 2006, more than 20 percent of Wangxiang's sales were for export, mainly to the U.S. market.

A key component of Wanxiang's business strategy has been upgrading (that is, steadily broadening and upgrading its product mix) combined with cost cutting. The strategy involves three interrelated efforts: (1) expanding the set of parts the company is capable of producing; (2) within each product category, creating more complex and demanding parts with higher value added; (3) entering more-lucrative markets, especially moving sales from the after-market to the original-equipment-assemblers market and from the domestic to the foreign market. A recent World Bank study on the pattern of innovation of Wangxiang found that much of its strategy for traditional auto parts involves incremental innovation (to cut costs) and acquisitions of existing firms, both domestic and international (to enter new markets). That is, most of its efforts were "new to firm" rather than "new to market." Nonetheless, the efforts were rational and effective. Wangxiang has implemented an increasingly active approach to the acquisition, absorption, and improvement of technology in recent years. In 1994, Wangxiang Qianchao, the group's publicly traded subsidiary, established its own R&D department. By 2006, the central R&D facilities employed 215 researchers, and R&D expenditures accounted for 4.5–5.0 percent of sales revenue—a share similar to the average for major global auto suppliers. A number of other group companies, including Wangxiang America, also established their own R&D facilities.

Source: World Bank, unpublished data.

land-intensive exports (sugar, rice, and bananas) to labor-intensive ones (textiles and shoes) and was led by small and medium-sized enterprises. Labor productivity in manufacturing grew at an annual rate of 6.77 percent during 1953–81 in Taiwan, China, while the ratio of R&D to GDP was merely 0.53 percent around the end of the 1970s (Hou and

Gee 1993, p. 384). Entering the automobile manufacturing industry turned out not to be part of the solution to the set of equations in Taiwan, China, although it was in the Republic of Korea. Chinese enterprises must also solve the set of equations to identify the right solution. Solving the set of equations leads to the next two issues.

Government vs. the Market

Government actively promotes innovation in virtually all countries. However, it remains crucial to realize that innovation is, after all, a business matter. Innovation, when it is broadly defined to include adaptation and adoption, is about the acquisition of certain technologies to which a firm does not have access. There are no obvious "silver bullet" answers to questions such as: What technology should be acquired—the higher one or the lower one? And how should it be acquired—make or buy? Decisions regarding those matters involve careful calculation of their effects on cost, revenue, and profitability based on forecasts of a range of parameters such as output and input prices, and they inevitably involve a process of trial and error. The best the government can do is to leave those decisions to business enterprises.

In a country like China, where the government has great power to mobilize resources and direct industries and enterprises, the need to allow enterprises to arrive at their own technology solutions as a business matter is particularly relevant. Catch-up is not about advances in a few high-tech sectors. It means the upgrading of the technological level and innovation capacity of a nation's industry as a whole. Aggressive government actions to accelerate the advances of a selected set of high-tech sectors would typically require a huge capital investment. Yet despite that investment, the industries and enterprises arising from it may fail to reflect China's comparative advantages and might require government protection to survive. In the meantime, those industries and enterprises that have received less government support may indeed better reflect China's comparative advantages (Lin 2007).

As Chinese enterprises enhance their innovation capabilities, a decline in their dependence on foreign technology is a natural result. The government is well advised to encourage Chinese enterprises to raise their capabilities for technology creation. But achieving the right balance between technological independence and openness means that the optimal level of independence from foreign technology is not the highest one but the one that contributes most to the development of technological capacity and ultimately to the sustainability of economic development. It takes the private market to find out and approach such an optimal level of technology independence. The central factor on which to focus is the absorptive capacity of domestically funded enterprises, which determines the speed with which they move on to technology creation. Again, caution is required to ensure that the reduction of dependence on foreign technology is treated as a means, not as the end.

R&D Expenditure vs. Other Factors of Innovation

China's emphasis on R&D investment must be at an appropriate level. Its spending on R&D as a share of GDP, 1.4 percent in 2006, is now the highest in the developing world, higher than India's and Brazil's. Although the measured share in China is still lower than the world average (1.6 percent) and that of developed countries (2.2 percent), China's expenditure statistics are not comparable to those for Organisation for Economic Co-operation and Development (OECD) countries because China does not attach values to tax incentives (expenditures), whereas OECD countries do. A further increase of R&D outlays to 2 percent of GDP and more over the longer term would be desirable. However, increased R&D expenditure leads to increased innovativeness only when it is matched with increases in other inputs, such as science and technology (S&T) employment and infrastructure. To the extent that the availability of other inputs may constrain the absorptive capacity for R&D, too rapid an increase in R&D risks misallocating resources (Yusuf, Wang, and Nabeshima 2009). A stronger emphasis on the effectiveness and efficiency of R&D spending, especially public R&D spending, is highly desirable.

Conclusions

China has achieved remarkable progress in industrialization and development. Since the late 1970s, continuous efforts in technology importation, absorption, adaptation, and creation have sustained the spectacular performance of the Chinese economy. Entering the 21st century, however, China is approaching a critical stage of its development. In the international market, Chinese enterprises are becoming perceived as strong competitors, face less favorable terms of technology importation, and have a diminished backlog of technology available for import. While those conditions suggest the need for a greater capability to achieve indigenous innovation, the capacity of Chinese enterprises for indigenous innovation appears inadequate to cope with the challenge.

At the domestic frontier, the sustainability of the growth model that China has followed over the past decades has been called into serious question because of its excessive reliance on capital and resources as opposed to knowledge and innovation. A transformation of China's economic growth strategy toward one that is based more on efficiency and knowledge is widely recognized as essential to China's long-term prosperity and has been at the center of the government's "scientific development strategy." In addition, Chinese enterprises are also expected to create jobs for the labor force, 80 percent of which does not have any education beyond the junior secondary level. In sum, Chinese enterprises must be not only competitive but also innovative and job creating.

The government of China has laid out a strategy of enterprise-led indigenous innovation in the 2006 S&T program. In view of the nature of the challenges Chinese enterprises are facing, a successful implementation of this strategy would require clarity concerning the objective of innovation. As important as it is, innovation is just a means, not the end. The ability of Chinese enterprises to promote innovation depends critically on the extent to which their efforts are oriented toward the larger objective of sustaining China's economic development. Serving the larger objective involves the pursuit of a balanced strategy in a number of dimensions.

The first is a proper balance between technological creation, on the one hand, and adaptation and adoption, on the other. As do many other developing countries, such as India, China stands to gain from a broad interpretation of innovation that includes (1) technology creation, adaptation, and adoption and (2) sustained efforts in promoting technology adaptation and adoption. The case is particularly strong in view of the pressure for job creation.

The second is the limit of government promotion of innovation to its proper sphere relative to market-based efforts. To ensure that the technologies employed in Chinese industry fully reflect China's comparative advantage, microeconomic decisions on technological innovation are better left to business enterprises and the market.

The third is the balance between R&D expenditure and other factors of innovation such as S&T employment and infrastructure, whose supply constrains the efficiency of R&D spending. As R&D outlays increase to reach 2 percent of GDP as targeted by the government's 11th Five-Year Plan (2006–2010), a sharper focus on the effectiveness and efficiency of R&D spending, particularly public R&D spending, is highly desirable.

CHAPTER 2

Creating the Right Incentives

The vision of the Chinese government regarding a technological innovation system emphasizes the "leading" role that should be played by enterprises and the "guiding" role that should be played by the market with regard to technological innovation. This vision, a reflection of the government's recognition of the fundamental weaknesses of the existing system, appears to be well-defined in the context of China's ongoing transition from a centrally planned economy to a market economy.

Indeed, in a market economy of private enterprises, technological innovation is inherently "enterprise led." In essence, "innovation is the specific instrument of entrepreneurship" (Baumol, Litan, and Schramm 2007; Drucker 1985, p. 3). Innovation can be seen as a marriage between new knowledge (as embodied in an invention) and entrepreneurship; the marriage is necessary to introduce the invention to the marketplace (Baumol, Litan, and Schramm 2007, p. 5).

As seen by Schumpeter, the process of innovation in a market economy takes two forms (Martin and Scott 1999). One is the never-ending cycle of entry by innovative small firms followed by the commercial application of new products or processes; the displacement of incumbents; and the entry of yet another generation of small, innovative firms— a process that may be characterized as one of "creative destruction." The second is routinized innovation through the engagement of large, established firms in risky, large-scale R&D activities, which makes innovation "a regular and even ordinary component" of the activities of the firm (Baumol 2002, p. 4).

Under each form of innovation, the market both spurs and guides firms. The market *spurs* firms to innovate through competition, which makes it difficult for them to otherwise maximize their value or even survive. In particular, under oligopolistic competition among large, high-technology firms, innovation replaces price as the primary competitive weapon (Baumol 2002, p. 4). Even in markets in which price remains the primary instrument of competition, innovation is an effective way to reduce cost.

Well-functioning markets also *guide* firms in the process of innovation, as well as in all other investment decisions, by providing economically meaningful prices that reflect resource constraints and consumer preferences. To a value-maximizing firm, investing in a technology that is new to it—through R&D activities, technology transfer, or in some other way—is just one among many investment options, all of which involve risks and uncertainties. The firm applies the same kind of calculations to all its perceived investment options to find those that will maximize its value. Obviously, such investment decisions make economic sense only when the market functions in a sound manner to generate economically meaningful prices.

China differs from most advanced market economies in that it had a long history as a central planning economy before embracing the notion of a free market economy of private enterprises.[6] In pre-reform China, the government took the place of entrepreneurship and the market. As was the case in most other centrally planned economies, technological innovation was led by the government and performed largely by government-run research and development institutions (RDIs) and higher-education institutions (HEIs) in a top-down manner. Pre-reform SOEs were little more than production arms of government departments. They produced what was required by government plans with the resources allocated to them by those plans. They had neither the motivation nor the autonomy to pursue value maximization, let alone technological innovation. Technological advances, if any, were mostly initiated and managed by the government through public RDIs and industry departments. SOEs typically played a more passive role in accepting and absorbing imported technologies and transforming R&D results into production capacity. Government decisions on technological advances, such as what to innovate and how to innovate, were not based on market prices because such prices did not exist.

China's market-oriented reforms and the opening up of its economy have moved it away from the government-led model of technological innovation toward one that is enterprise led and market based. However, the transition is still quite far from completion. A system of technological innovation "led by enterprises and guided by the market" has not emerged yet (MOST Study Team 2006, p. 15). Hence, the further strengthening of the role of business sector enterprises, including SOEs, and of the market are fundamentally important to China's long-term capacity to innovate.

The lack of an enterprise-led and market-guided technological innovation system makes the challenge of technological innovation quite different for China than it is for more advanced market economies. For the latter, discussions of innovation policy can take value-maximizing firms and well-functioning markets more or less for granted and concentrate on the next level of issues such as R&D intensity, S&T human resources, tax incentives, and so on. Those issues are of course also relevant to China. However, China has to face the more fundamental challenge first: to create the microinstitutional foundation for innovation by putting in place value-maximizing enterprises and well-functioning markets. Even for advanced market economies, "it would be naive to assume that patent protection and R&D subsidies would be sufficient to foster innovation and productivity growth" (Aghion 2006). But that is particularly true for China. If the challenges discussed in chapter 1 can be labeled "development challenges," the creation of the microinstitutional foundation for innovation is more in the nature of a "transition challenge."

Who Performs R&D in China?

To what extent do enterprises now lead China's innovation activity? What kinds of enterprises—large, small, state-owned, private—are active in R&D? These are important questions for an understanding of where China stands in the transition to enterprise-led, market-based innovation.

The Shift from RDIs and HEIs to Enterprises

Over the past decade or so, business enterprises have replaced RDIs and HEIs, which are largely government owned, to become the most important sector for R&D. In 1995, RDIs and HEIs jointly accounted for 54.1 percent of China's total R&D expenditure and 51.7 percent of the full-time equivalent of R&D personnel. By 2006, those shares had dropped to 28.1 percent and 31.6 percent, respectively. In contrast, 71.1 percent of total R&D expenditure was spent by enterprises in 2006, which also employed 65.8 percent of the country's R&D personnel.[7]

In addition to the increasing R&D efforts of business enterprises, continuous efforts in reforming RDIs since the mid-1980s (box 2.1) has made

Box 2.1

China's Reform of RDIs

In pre-reform China, RDIs were established by all levels of government and managed as public service units (PSUs).[a] PSUs are funded and run by the government, and PSU workers are public sector employees. In the pre-reform era, each RDI was affiliated with (1) the China Academy of Science, (2) HEIs, (3) industrial ministries and bureaus, (4) local governments, or (5) the military.

The reform of RDIs started in 1985. A primary objective was to facilitate the integration of RDIs with economic activity by making most RDI work increasingly demand driven. Each RDI was classified into one of three types of research: (1) basic, (2) public-benefit related, or (3) development. Financial support of development research units was gradually reduced over the period from 1985 to 1990, forcing them to obtain supplementary funding from the market.

In 1995, the government decided to further "liberalize" some RDIs by transforming them into business enterprises. Implementation of this new strategy intensified when, in 1998, the government abolished 10 industrial ministries with which a large number of RDIs were affiliated.

Some 376 RDIs were formally transformed into business companies in 1999 and 2000; others, although not formally transformed, are operating as partially or completely self-financing entities. The conversion process has created some successful innovative enterprises. One example is the China Academy of Telecommunications Technology created in the 1950s as a PSU affiliated with the Ministry of Post and Telecommunication.[b] In 1998, the academy established its flagship business company, Datang Telecommunication, and conducted an initial public offering of stock (IPO) in that company. The academy itself was transformed into a business holding company, with its business arms organized into a conglomerate known as the Datang Telecom Technology and Industry Group. Datang is now a large SOE group in the portfolio of the State-Owned Assets Supervision and Administration Commission (SASAC) and is leading the development and commercialization of China's own version of 3G technology for mobile phones.[c]

(continued)

Box 2.1 *(Continued)*

Also in the late 1990s, the government identified a group of RDIs that it said could not be expected to be self-sufficient. For those RDIs, the government piloted a management model in which they would operate as nonprofit organizations (NPOs). In exchange for the funding of research and facilities, the government required the nonprofit RDIs to restructure themselves and lay off redundant employees. Implementation of this reform started in November 2001, when 98 RDIs under four ministries were selected as pilot NPOs. A second group of 107 institutions under nine ministries joined the pilot in October 2002. Of the 205 RDIs in those two groups, 142 received final approval as NPOs.

a. More details on PSUs and their reform are in World Bank (2005).
b. http://www.catt.ac.cn/english/portfolio.asp.
c. The technology, called TD-SCDMA (time division-synchronous code division multiple access), is being jointly pursued by the China Academy of Telecommunications Technology (CATT), Datang, and Siemens AG. Further information is available at http://en.wikipedia.org/wiki/TD-SCDMA.

them much more market-oriented, which resulted in a transformation of some of them into business companies. To what extent has the transformation of RDIs contributed to the observed increase in the share of enterprises in R&D expenditure? Lack of R&D data from the transformed RDIs makes it difficult to see precisely their share in the total for Chinese enterprises; but some data from 2001 and 2002 on their "S&T development expenditure," of which R&D is a subset, suggest that the contribution has been minor—on the order of about 2 percent.[8]

The Dominant Role of SOEs

Business enterprises overall have replaced RDIs and HEIs to become the most important performing sector of R&D in China, but the role of small enterprises remains limited and has not seen significant change over the past few years (figures 2.1 and 2.2),[9] and the shares of domestic private LMEs are smaller still (figure 2.3). Thus, it is LMEs owned and controlled by the state (SOE/LMEs) that now conduct the largest portion of R&D in China.[10]

Innovation may take two forms in a Schumpeterian sense. Whether a particular R&D activity is better undertaken by a large firm or a small one depends in principle on the technical and economic nature of that activity. But in China, the dominant share of LMEs in R&D suggests that the innovativeness of large firms is critical. How could China increase

Figure 2.1   Distribution of China's R&D Expenditures, by Performing Sector, 2000–06

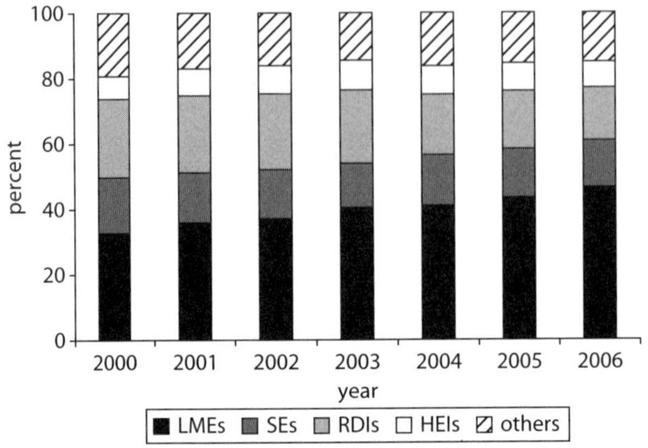

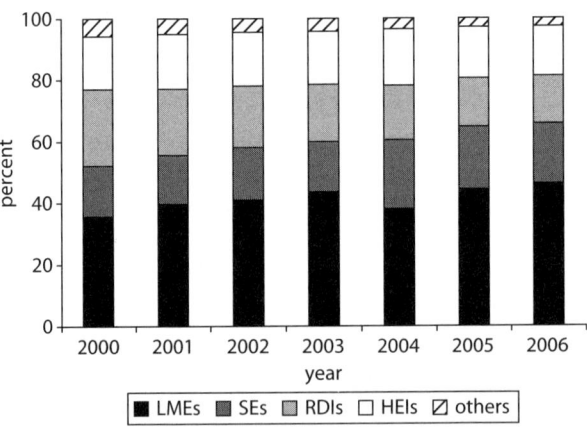

Source: NBS and MOST 2007, table 1-9, p. 7.

Figure 2.2   Distribution of the Full-time Equivalent (FTE) of China's R&D Personnel, by Performing Sector, 2000–06

Source: NBS and MOST 2007, table 1-5, p. 5.

the number of large, innovative firms? The picture of corporate China shown in figure 2.3 implies three parallel routes to that goal: (1) increasing the size of private firms, (2) privatizing SOE/LMEs, and (3) making existing SOE/LMEs more innovative. Because the growth of private firms takes time and the privatization of SOE/LMEs has to be gradual, much of China's enterprise-led innovation in the near future will have

Figure 2.3   Distribution of China's R&D Effort, by Type of Performer and, for LMEs, by Type of Ownership, 2006

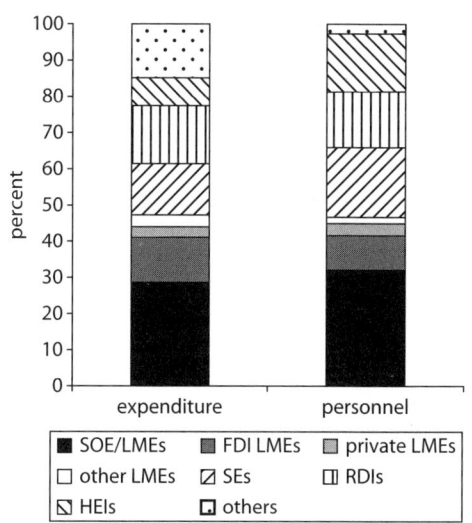

Source: NBS and MOST 2007, chapter 3.

to be led by existing SOE/LMEs. Their commitment to, and effectiveness in, innovation are, therefore, of great importance.

Making SOEs More Innovative

The R&D commitment and effectiveness of China's large SOEs do not appear to be adequate. First, the *commitment* of China's SOE/LMEs to technical innovation seems low when evaluated on an international scale. In 2006, they spent, on average, nearly 1 percent of their sales revenue on R&D. Although that share was significantly higher than that of China's domestic private enterprises (table 2.1), it is inadequate when compared with that of potential competitors in the international market. For example, 2005 R&D expenditures represented 4 percent of sales revenue at Toyota and 5 percent at Ford, 15 percent at pharmaceuticals giant Pfizer, and 16 percent at software leader Microsoft.[11] The State-Owned Assets Supervision and Administration Commission (SASAC) of the State Counci,[12] the ownership agency of the central government that acts as the shareholder in SOEs, has expressed its concern over the relatively low commitment demonstrated by such comparisons, characterizing the problems with its portfolio enterprises as "manufacturing without innovation."[13]

Table 2.1   R&D Expenditures and Performance of LMEs in China, by Type of Ownership, 2006

Type of LME ownership	Number of enterprises	R&D expenditure as percentage of sales revenue of core businesses	Scientists and engineers as percentage of total employment	Patent applications per million yuan of R&D expenditure	Patents owned per 100 scientists and engineers
Total, including	32,647	0.80	2.73	4.09	2.50
State owned and controlled	12,954	0.92	3.63	3.67	1.79
Privately owned	6,755	0.50	1.75	10.15	5.23
With investments from Hong Kong, Macau, and Taiwan, China	5,458	0.70	1.28	6.36	3.84
With foreign investments	6,128	0.60	2.03	3.08	3.47

Source: NBS and MOST 2007, p. 101.

In terms of R&D *effectiveness*, that is, results for a given level of R&D expenditure and employment, SOE/LMEs appear to be less effective than domestic private LMEs. The latter seem to be able to file more patent applications per million yuan of R&D expenditure, and they own more patents per 100 scientists and engineers employed, than do their SOE counterparts (table 2.1).[14] A larger study, conducted jointly by the OECD and NBS on firm-level data of 140,000–180,000 industrial enterprises for the 1998–2003 period, also points to the low productivity of SOEs in comparison with private firms: "On the basis of a value added measure of output, total factor productivity (TFP) in private sector companies, after taking into account the impact of firm size, location, and industry, is double that in directly state controlled firms (90 percent to 123 percent higher). Reforms that have changed the nature of state control over enterprises, by allowing control to be exercised indirectly—through other companies—have boosted productivity. These indirectly state controlled firms are about 50 percent more productive" (Dougherty, Herd, and He 2007, p. 318).

Because it will take some time for the domestic private sector to grow enough to take over the leading role of SOE/LMEs in innovation, a dual approach seems the best strategy for China to follow in the near future: (1) further strengthen the corporate governance of SOEs to improve their innovation performance as much as possible and (2) further downsize the scale of state ownership in business enterprises.

Further Reform of Corporate Governance

How can corporate governance be relevant to innovation? The answer is that innovation—the acquisition of a technology that is new to a firm—is just one of the many business activities a firm undertakes. Sound corporate governance affects innovation in the same way that it affects all other business activities, by ensuring that the right incentives are at work.

The traditional system of SOE governance tends to retard innovation for several reasons. First, if SOEs pay attention to economic efficiency at all, it is still only one of many objectives they are expected to pursue. Hence, when SOEs engage in innovation, they often do so "following instructions from above," with little regard for market conditions. As one recent study by a well-informed researcher has pointed out, innovation in SOEs is often prize oriented instead of market oriented (Zhang 2007a).

Second, many SOEs still lack a functioning board of directors, which in the private sector constitutes the business-oriented decision-making center of the firm. Hence, business decisions, including those involving innovation, have to be made either by government officials outside the enterprise or by government-supervised managers, who do not always seek to maximize the value of the firm.

Third, a lack of flexibility and adaptability in the management of human resources, particularly in performance evaluation and compensation, tends to compromise innovation activities in SOEs. These three areas should receive adequate attention in further reform of corporate governance.

Strengthening the focus on long-term productivity enhancement. Since the establishment of the central and local SASACs in 2003, China has made remarkable progress in reforming SOE governance. The SASACs' insistence on improving the value of state assets has dramatically strengthened the profit orientation of SOEs and reduced their attention to nonbusiness objectives. In particular, in the absence of genuine board governance, central SASAC set up a performance evaluation system for top managers and directors at centrally owned SOEs.

The performance evaluation system is expected to be further enhanced by 2010, when economic value added (EVA) is incorporated as a central instrument. EVA emphasizes the opportunity cost of capital, a concept that, until recently, has been largely ignored in China. Measuring value creation after accounting for the opportunity cost of capital tends to spur SOEs to focus more on productivity as opposed to growth in revenues, profits, and assets. Focusing on productivity naturally leads to a focus on technological absorption, adaptation, and creation.

However, innovation often involves long-term investment and repeated failures. Therefore, SASACs would be well advised to ensure that the EVA-based system is carefully designed to adequately account for the potential for long-term value generated by innovation activities, even though not all of those activities will be successful. If the evaluation system is biased in favor of a short-term return to capital, it may actually discourage risk taking. This concern is particularly important for SOEs that are not listed on a stock exchange. Under the current evaluation system, SASACs have already started giving more weight to innovation (increases in R&D spending, new patent applications, technology transfer, and so on). That emphasis would need to be incorporated in an EVA-based system.

Pursuing board governance. The extent to which such administrative reviews can encourage innovation, even with an EVA-based instrument, is far from clear. A surer approach would be to continue the development of SOE boards of directors consisting of independent and broadly experienced members who can define long-term goals and champion and assess efforts at long-term innovation.

After its establishment in 2003, central SASAC quickly launched an experiment to bring board governance to its SOEs.[15] The motivation for the experiment was that, without functioning board governance in its SOEs, central SASAC would be forced to centralize business decision-making to protect the state's ownership interests. Although progress has been slow—at the end of 2007, the experiment covered only 19 of 153 central SOEs—board governance should continue to be a priority for reform. Because a business governance culture characterized by a board that includes external, independent directors is new in China, its spread will inevitably take time. Stronger commitment and political determination from the leadership of governments at all levels are crucial for the reform to move forward. As progress is made in introducing board governance, it is advisable for the government, in particular, SASACs to delegate more decision-making power to boards of directors.

Improving human resources management. Human resources management, including compensation, is critical to innovativeness. The inflexibility of state control over performance evaluation and compensation has provided little incentive for long-term innovation by China's large firms. Because the payback from innovation is often uncertain and long term, rewarding managers with equity ownership, especially through stock options, is frequently the most effective means of aligning the interests of management and owners. However, only a small fraction of China's SOEs are publicly listed, so opportunities for such incentive-based compensation are limited.

In the absence of meaningful equity incentives, performance incentives tend to emphasize bonuses, which currently are limited by administrative constraints and gaps in corporate governance. It seems clear that centralized management of performance incentives by SASACs can hardly provide the needed flexibility and adaptability to encourage innovativeness without running the risk of losing control over management pay. So long as the companies remain under state control, the way out is board governance that decentralizes information collection, performance evaluation, and compensation decisions.

Further Scale-down of State Ownership
Continued reform in SOE governance is surely important to China, but the improvement of corporate governance alone is not likely to make

Chinese SOEs as innovative as their private counterparts. Corporate governance reform in fields such as board governance and management pay in China's SOE sector will inevitably meet great difficulties, which set limits on how much can be changed in a short time. In addition, China's SOE sector is growing fast. Despite a sharp downsizing of employment at nonfinancial SOEs and a large reduction in the number of such firms, their output and assets increased 118 percent and 80 percent, respectively, from 1998 to 2005. That growth suggests that SOEs are the recipients of a continuous injection of economic resources. Given the productivity gap between SOEs and non-SOEs found in the OECD-NBS joint study, the shift of resources to SOEs implies a significant reduction in China's overall economic efficiency.

One important channel for public resources to be pumped into SOEs is related to the government's dividend policy, or indeed the absence of dividend policy. Although parent companies of SOE groups collect dividends from their subsidiaries, the SOE sector as a whole did not pay any dividends to the government between 1994 and 2007. The retention of all after-tax profit has become significant with the enormous rise of SOE profitability in recent years. From 1998 to 2007, the total pretax profit of all nonfinancial SOEs rose from Y 21.3 billion in 1998, or 0.3 percent of GDP, to Y 1.62 trillion, or 6.6 percent of GDP.[16]

In 2008, the government began collecting dividends from wholly state-owned companies on an experimental basis. One of three dividend rates is applied to each firm—10 percent, 5 percent, or 0 percent. The move is a good beginning, but as experience is accumulated, two changes are desirable: First, the government should eliminate the 0 percent rate over one or two years. Second, it should switch from a uniform set of positive rates to variable rates governed in part by the potential of each SOE to innovate and grow—and the judgment about such potential cannot be accurately rendered from afar but rather only in a decentralized way, by boards of directors with decision-making power. A uniform dividend rate leaves too much free cash to firms that do not have much potential for innovation and growth and too little to those that do. In particular, given the limited role that the Chinese capital market can play in financing R&D, SOEs that engage in high-risk innovation activities may run into financial trouble when a uniform dividend rate becomes high. Again, having boards of directors with decision-making power is critical to a decentralized determination of the dividend rate.[17]

In addition to implementation of the dividend policy, the government could take further action to dilute state control and ownership by pushing

SOEs into the capital market through the issuance of secondary shares to private investors. Such issuance can serve two purposes: (1) enable SOEs to expand their R&D efforts and (2) increase the exposure of the firm to the discipline of the capital markets and governance norms. The state's share of ownership in SOEs that are listed on a stock exchange averages around 70 percent; the issuance of secondary shares that dilute state ownership to 33 percent could raise substantial funds suitable for long-term and risky investments in R&D. Having up to 67 percent of shares owned by private investors with a long-term view should make it easier for an experienced, professional board of directors to support an innovation program developed and implemented by management. Freedom from government constraints should allow the firm to provide managers with market-competitive compensation that includes substantial incentives for innovation.

Strengthening External Incentives

The effort to create the right incentives to innovate goes beyond issues of governance and ownership. Indeed, "manufacturing without innovation" is a symptom of most Chinese private enterprises as well as of SOEs. In a recent study, the All China Federation of Industry and Commerce (the Chinese chamber of commerce) found "most" private enterprises "not interested in" or "unwilling to engage in" innovation: "There are over 4.6 million private firms in China, only a tiny number of them engage in technological innovation. Even among the 150,000 private S&T enterprises, only a small fraction has genuine achievements of technological innovation, not to mention few of them have invention patents and original technology."[18] The report attributes these results to two factors. First, some private enterprises are still struggling to identify their core business so as to survive in the marketplace, and therefore they are not in the right mode to engage in technological investigations. Second, some others are tempted by quick profits and do not see innovation as the best way to make money: "When there exist in the market opportunities of windfall profit, such as real estate development and resources extraction, most private enterprises, including even a substantial portion of private S&T enterprises, are not willing to invest their time and money in genuine technological innovation."

The federation's conclusions seem to represent the consensus view among Chinese policy makers and advisors, who often argue that at least a large portion of private enterprises have little incentive to invest in innovation and technological progress.[19] Their natural tendency is to maximize short-term profit through the expansion of their existing production capacity and market share as well by controlling costs.

Getting Market Incentives Right

For business enterprises, innovation is of course only a means, not an end. With sound governance and ownership, business enterprises do not engage in innovation unless it promises to help maximize the value of the firm, that is, to maximize the present value of its expected future stream of profits. In other words, enterprises make decisions on innovation—such as to innovate or not to innovate, what to innovate and how much, how to innovate—to maximize their profit. Therefore, creating a market environment that provides the right incentives for firms to innovate is an essential complement to the creation of value-maximizing enterprises through the reform of their ownership and governance structures. In that way, the reformed firms' heightened motivation for profit can be translated into a strong motive for innovation. Creating the right market incentives is likely to involve reforms in many areas, especially pricing, competition, and the market for corporate control.

First, having the right market incentives means that market prices must fully reflect resource constraints and consumer preferences. All investment decisions are made on the basis of prevailing prices for all outputs and inputs. Distorted prices distort economic calculations and mislead investment decisions.[20] In particular, the underpricing of inputs such as energy, natural resources, land, environment quality (in the form of weak environmental regulations and enforcement), and labor (in the form of weak protection of labor rights) has the potential to discourage investment in technological progress that is more energy efficient, resource saving, or beneficial to labor. Despite nearly three decades of market liberalization, prices for land, energy, water, and mineral resources in China remain controlled or heavily influenced by the government. At least a significant portion of the prices of those resources is too low to fully reflect their true social cost, including their scarcity. Despite a recent strengthening trend, environmental regulations are often poorly enforced in the effort to prop up GDP and protect jobs.

Likewise, the protection of labor rights is weak, particularly in small and medium-sized private enterprises operating in medium- or low-tech sectors; the problem reflects the pressure of excess low-skilled labor and the absence of organized bargaining power on the part of workers. The National Development and Reform Commission (NDRC) announced at the end of 2007 that it would reform the pricing of resource-related products and institute fees related to environmental protection despite rising inflation pressures.[21] The enactment of the Labor Contract Law at the beginning of 2008 is also expected to substantially strengthen the protection of labor rights. Continuing reforms along these lines are likely to lead to positive changes in market incentives facing business enterprises.

Second, having the right market incentives means that enterprises must be exposed to the pressure of competition. The power of the market in spurring enterprises to innovate is substantially weakened when inefficient producers can be sheltered from competition, such as through barriers to entry and exit and inadequate antitrust regulation. In China, a few state-owned conglomerates continue to enjoy monopolistic advantages granted by the government. However, the reach of the Anti-Monopoly Law promulgated by the National People's Congress in August 2007 includes monopolistic SOEs. The law's article 7 requires the state to regulate the operations of such SOEs to "protect consumers' interest and promote technological progress." The government could consider implementing this law in the SOE sector through a special regulation. The government could also be more mindful of the potential for its industrial policies to hamper competition (box 2.2).

Box 2.2

Entry Barriers Created by Industry Policies for the Dairy Industry

On June 4, 2008, the NDRC enacted a regulation to implement industrial policies in the dairy industry. The regulation requires new entrants to meet the following criteria before their investment projects will be "reviewed and permitted (*hezhun*)": (1) the size of the investment is no less than Y 30 million–50 million; (2) the distance of the new project from existing dairy production firms is no less than 100 kilometers (km) in northern China and 60 km in southern China; (3) the project's daily capacity for processing liquid milk is, in the north, no less than 500 tons for new facilities and 300 tons for the expansion of existing facilities; and 200 tons and 100 tons, respectively, in the south; (4) the project has a stable source of milk supply; (5) the existing net worth of the investor is no less than twice the equity investment needed by the new project; (6) existing total assets of the investor are no less than three times the total investment of the new project; (7) the investor's ratio of debt to assets is no higher than 0.7; and (8) the investor has been profitable for three consecutive years.

Source: http://www.ndrc.gov.cn/zcfb/zcfbgg/2008gonggao/t20080604_216116.htm.

Finally, having the right market incentives means encouraging merger and acquisition so as to rationalize the industrial organization. A widely accepted principle is that an oligopolistic market structure provides more of an incentive to innovate than does either full competition or monopoly: full competition tends to reduce the appropriability of innovative firms because of the spillover problem, and monopoly provides weak incentives to invest in innovation (Baumol 2002, p. 45). China's current structure of industrial organization suffers from the weaknesses of both extremes. A small number of SOE conglomerates have received government grants of monopolistic advantage, and the major part of remaining Chinese industries are inadequately concentrated, mainly because of so many small producers equipped only with simple, cheap, and inefficient technologies. Such conditions reflect the rapid pace of business creation over recent decades (more details are in chapter 3), the limited access to finance for SMEs, and the inherently long time it takes for SMEs to grow to efficient scale. However, policies designed to encourage merger and acquisition could help accelerate the process of industrial rationalization, which could in turn strengthen the incentives to innovate.

Making Good Use of Fiscal Incentives
As a result of inevitable spillovers, the social return to R&D investment—which can be as high as 100 percent—is typically much higher than the private return, which is often in the neighborhood of 28 percent.[22] This discrepancy, resulting from market failure, justifies government interventions of various sorts. Fiscal incentives are an important and commonly used type of such intervention.

China has instituted a wide range of fiscal incentives, including tax incentives for R&D spending and for hi- and new-tech development zones, and direct grants to specific R&D activities. In particular, the current tax incentives are generous by international and East Asian standards. They include an exemption of up to 150 percent of R&D expenditure from corporate income tax and the provision of carrying forward any unused amount to offset tax liabilities up to four years in the future. Accelerated depreciation allowances permit firms to treat expenditures on equipment worth less than Y 300,000 as overhead; for more-expensive equipment, the depreciation period can be shortened to as little as three years. The revised Corporate Income Tax Law (effective on January 1, 2008) states in Article 28 that a 15 percent tax rate applies to "hi- and new-tech firms," compared with the normal rate of 25 percent. Article 93 of the Corporate Income Tax Law Enforcement Rules further defines criteria of "hi- and new-tech firms" for tax purposes and states that a regulation concerning the identification of "hi- and new-tech firms" will be formulated jointly by ministries of S&T, finance, and tax administration. Companies that incur heavy expenditure on fixed investment as a part of their R&D activities will benefit from a switch to a consumption-type value added tax (VAT), which has been under pilot implementation.[23] Exemption from import duties on equipment for R&D further augments earnings. Firms in the biotech, telecom, new materials, aeronautics, information technology, and electronics fields derive substantial benefit from such preferential tax treatment. Tax incentives are complemented by direct central and subnational government spending on R&D. Grants by various ministries have reached significant levels and are rising at a fast rate.

The important question now for China regarding fiscal incentives is not how to make them more generous, but how to tailor them to produce the best results, because increased spending on activities classified for tax purposes as R&D might have low social returns. Low returns are more likely in circumstances in which firms are still mainly in the assimilation stage and poorly equipped in terms of strategy, managerial expertise, organizational design, and technical skills to conduct meaningful research or to use research findings for commercial purposes. These constraints, especially the shortage of seasoned midlevel research managers, might argue for a design of tax incentives that encourages companies to pool their research efforts and form a variety of alliances. The formation of research consortia and joint programs with local or foreign HEIs are good approaches to consider. Thus, tax incentives could be made particularly generous for joint research programs with foreign companies based on the scale of the foreign involvement and the industrial sector that is the focus of the research. This approach would encourage multinational companies that already benefit from incentives to localize research activities and work more closely with Chinese firms.[24]

Incentives to "offshore" some research and engage more closely with researchers abroad would recognize the realities of a globalizing research environment. The value of offshoring does not undermine the case for strengthening local capacity, but it does argue for taking full advantage of international research capabilities, where possible, in the interests of enhancing competitiveness. Offshoring research could also put pressure on local research entities to improve their own performance, and international joint research ventures can also be an important vehicle for technology transfer.

In short, as circumstances permit, the fiscally cost-effective approach to supporting corporate research at China's current stage of development might be one that stresses both locally and globally pooled efforts. This approach would recognize that in certain cases it might be more efficient to allow Chinese researchers to continue working abroad rather than offering them generous incentives to return to what might be initially a less productive niche in the local research environment. The approach would also benefit from a further strengthening of the institutions protecting intellectual property rights (IPR), especially the courts.

Raising the Demand for Innovation

In addition to government actions that stimulate the supply of innovation, experience from Japan, western European countries, and some other OECD members indicates that governments can do a great deal to increase the demand for innovation. Procurement and standard setting are two of the many tools that governments can use to achieve that goal.

Government Procurement

Government procurement is probably an underused instrument for encouraging innovation in China. According to estimates in an OECD (2002) study, the 1998 ratio of procurement to GDP was 20.0 percent for all levels of government for all OECD countries, and it was 14.5 percent for non-OECD countries. Although the data may not be strictly comparable, China's government procurement appears to be much smaller, representing only 1.6 percent of GDP in 2005 (table 2.2). However, the amount of government procurement in China has been growing rapidly, reaching an annual rate of 37 percent in 2005. And despite its small size in relation to GDP, the government market in absolute amount can be significant for some sectors in China. For instance, in 2005, China's government spent Y 20.4 billion on transportation equipment, which is equivalent to 2 percent of the total sales of the auto industry (DIE/DRC and others 2008; Zhang 2007b).

Table 2.2   Government Procurement in China, 2003–05

Indicator	2003	2004	2005
Total amount (Y billion)	165.9	213.6	292.8
Annual growth (percent)	64	29	37
Percentage of GDP	1.4	2.0	1.6

Source: Ministry of Finance, National Government Procurement Statistics, http://www.ccgp.gov.cn/tjzl/index.htm.

Indeed, managing government procurement is a relatively new topic in China. The first national guideline for government procurement was issued in 1999, and the Government Procurement Law was adopted by the National People's Congress in 2002. Despite the newness of the approach, however, the government's determination to support innovation through procurement has been made firm and clear. Following the adoption of the 2006 S&T program, a State Council circular announced policy directives for government procurement to support indigenous innovation.[25] It requires the establishment of a regime for government procurement of products of indigenous innovation; improving evaluation methods to give preferential treatment to products of indigenous innovation; and creating rules on government procurement of the first batch of indigenous innovation products and the ordering of goods and services that are of the nature of indigenous innovation. National defense procurement is also required to support indigenous innovation (Zhang 2007b).

In that spirit, the Ministry of Finance (MOF), in collaboration with the Ministry of Science and Technology (MOST) and the NDRC, released another set of regulations in 2007 and 2008 to implement the State Council policy directives. The key components of the policy framework established by these regulations include the following: (1) a "Catalog of Indigenous Innovation Products" to be compiled jointly by MOST, NDRC, and MOF; (2) seven criteria and four steps for the identification of indigenous innovation products;[26] (3) preferential treatment for products in the catalog (depending on procurement methods, the preferential treatment implies roughly a premium of 4–10 percent of the bidding price);[27] (4) procurement by the government of the first batch of indigenous innovation products that are not ready to sell in the market (*shougou*);[28] and (5) awarding of an R&D contract in an open and competitive manner when the government wishes to mobilize R&D activities for a critical innovation product, technology, or software (*dinggou*).[29]

Government procurement can help or hurt innovation (box 2.3). The key to success lies in open competition, as indicated by case studies of OECD countries. The government of China is still in the early stages of implementing innovation-supporting procurement policies. One may anticipate a number of issues that may require further policy action down the road.

The first issue is the risk of turning government procurement from an innovation-supporting instrument into one that protects national and local products from international and national competition. In particular, the long-standing problem of local protection may find a new form in government procurement (box 2.4).

Box 2.3

Government Procurement Practices: Hurting or Helping Innovation?

Competitiveness can suffer if government purchases become a guaranteed market. According to Michael Porter, "The German Bundespost, the state-owned telecommunications monopoly, is a notorious example." In the United States, "buy American" laws have affected some types of government procurement. Similar laws, or de facto exclusions of foreign suppliers, have been common in other countries as well. "The government market becomes the focus of attention, and domestic firms lobby for unusual product standards or other regulations to freeze out international rivals." When foreign suppliers are excluded from government procurement, "the result in most industries is that innovation and upgrading by domestic firms slow down.... Domestic firms are then unable to compete in international markets, and even more blatant favoritism at home becomes necessary to support them."

But government procurement can help innovation under the following circumstances:

- *Providing early demand.* Governments should provide early demand for advanced new products or services, pushing local suppliers into new areas.
- *Being a demanding and sophisticated buyer.* "Government agencies should set stringent product specifications and seek sophisticated product varieties rather than merely offer what domestic suppliers offer."
- *Setting internationally applicable product requirements.* "Government specifications should be set with an eye to what will be valued in other advanced nations, rather than reflecting only the nation's idiosyncratic needs."
- *Stimulating competition.* "Government procurement must include a strong element of competition if it is to upgrade the local industry." For example, in Japan, acting in its role as government buyer, NTT has typically ordered the next-generation systems rather than what Japanese suppliers currently produced. Most importantly, NTT has maintained a number of suppliers for each product, ensuring domestic competition for its business. "Foreign vendors must be allowed at least some access . . . to stimulate further innovation by domestic firms.... If domestic firms are weak, the best solution is to award foreign vendors some business and force domestic firms to upgrade their positions against a timetable" to retain some share of the government procurement market. "Shutting foreign firms out altogether and guaranteeing domestic firms the business will most likely mean that domestic firms will remain domestic."

Source: Porter 1990 , pp. 644–46.

Box 2.4

Using Government Procurement to Protect Local Production: The Case of Li Jiating of Yunnan Province

Li Jiating was formerly the governor of Yunnan province, whose capital city is Kunming. He resigned his position on June 1, 2001, and on May 9, 2003, a Beijing court found him guilty of accepting bribes. According to a news article, "About the downfall of Li Jiating, the feelings of workers of the Kunming Iron and Steel Company (KISC) are perhaps the most mixed. This company of tens of thousands employees, the largest enterprise in Yunnan province, was on the brink of bankruptcy when the stadium and sites of 'Kunming World Horticulture Exhibition' started construction in 1999. Li Jiating moved quickly to order that all steel needed by the exhibition-related construction must be procured from KISC, which turned that company around and saved tens of thousands of workers from the hardship of bankruptcy. When the national government launched the 'Western Development' program, Li Jiating again issued an explicit policy that all the 'top ten development projects' of Yunnan province must direct their steel orders to KISC. However, after June 1, 2001, when Li resigned, sales of KISC collapsed, with 50,000 tons of steel piling up in inventory in three months."

Source: http://www.gl.gxnews.com.cn/news/20030729/jctj/180439.htm.

The second issue one may anticipate is the challenge of following the procedures laid out to identify the "indigenous innovation products" for the catalog. Processes like that could be easily hijacked to become one more field of rent seeking. One potential way to guard against that risk is to focus more on observable and verifiable features of the innovativeness of the technology employed in production (Zhang 2007b).

Third, it is not clear how the upcoming catalog will be formulated to make the government a demanding buyer of technologically sophisticated products rather than merely a passive taker of what domestic suppliers offer.

Standard Setting

Government standard setting is another underused instrument to create demand in China for innovation. Standard setting allows governments and other entities to generate demand for advances in, for example, the performance, safety, energy efficiency, and environmental impact of products. Requirements for firms to improve products will pressure them to upgrade their own technology and performance. One major study finds, for instance, that the establishment of tough quality standards for exports by Japan in the 1950s and 1960s stimulated improvements in Japanese industry; the study finds that Germany had a similar experience (Porter 1990).[30] Also "particularly beneficial are stringent regulations that anticipate standards that will spread internationally." Thus, for instance, Sweden's tough standards for product safety and environmental protection "have been a significant source of competitive advantage in a variety of industries." In Japan, standards for energy use set by the Energy Conservation Law of 1979 led to many product improvements in air conditioners, refrigerators, and automobiles that enhanced the position of Japanese exporters. High standards have also encouraged the start of internationally competitive, specialized manufacturers and service firms. U.S. firms initially led in the export of pollution control equipment and services as a result of domestic environmental standards. With the development of more stringent standards in Denmark, Germany, and Sweden, however, firms from those countries began to gain global market share (Porter 1990, p. 648).

Firms may tend to focus more on the short-term costs of implementing standards than on the longer-term benefits for innovation, especially if a lack of such standards elsewhere gives foreign competitors a cost advantage. That attitude reflects a lack of understanding of how to create and sustain competitive advantage, according to Porter (1990): "Selling poorly performing, unsafe, or environmentally damaging products is not a route to real competitive advantage in sophisticated industries . . . especially in a world where environmental sensitivity and concern for social welfare are rising in all advanced nations. Sophisticated buyers will usually appreciate safer, cleaner, quieter products before governments do." Indeed, firms that are able to produce and sell more-sophisticated products in foreign markets may actually be able to gain competitive advantage by accelerating the implementation of tougher standards in those foreign markets (pp. 648–49). That is often the reason that exports to developed economies may help firms in developing economies become more innovative, as was the case in the Republic of Korea and in Taiwan, China (Hou and Gee 1993; Kim 1993).

China adopted its Standardization Law in 1989, which was followed by implementing regulations in 1990. The law establishes four levels of standards—national, trade, local, and enterprise. Except for enterprise-level standards, the government has the sole authority to set and modify standards. Authorized by the State Council (under the administration of the General Administration of Quality Supervision, Inspection, and Quarantine), a Standardization Administration China (SAC) is in charge of the unified administration of standards throughout China.

Overall, China is moving toward a more market-based system of standard setting with the greater participation of enterprises and a better balance among competing objectives (Zhang 2007b). However, China's approach to standards development has not done enough to support innovation. Case studies suggest considerable variation in China with regard to objectives and implementation patterns. For instance, an assessment of the electronics industry finds that some standard setting has been motivated by a desire to avoid royalty payments that protect domestic industry rather than to promote innovation and product improvements.[31] However, the SAC has recently announced a set of 12 measures to support indigenous innovation, including efforts to enable China's leading enterprises to participate in the modification of national and international standards.[32]

Outmoded standards or, even worse, use of standards for motives other than to promote innovation, would be a disservice to China. An analysis by Porter (1990) of the OECD experience shows that "regulation undermines competitive advantage . . . if a nation's regulations lag behind those of other nations or are anachronistic. Such regulations will retard innovation or channel the innovations of domestic firms in the wrong directions." For instance, limits on biotechnology research are considered to have threatened Germany's agrochemicals and pharmaceuticals sectors. As for competitiveness in export markets, "the practice of using idiosyncratic local regulations to protect a domestic industry will only work to ensure that its competitive success is domestic" (p. 649).

Standard setting needs to be efficient. If the process becomes protracted, and basic technological parameters remain in doubt, innovation slows down. To this end, there is a need to ensure that enterprise participation is done in a productive way. "In the United States and often in Europe, the process of reaching technical standards is frequently protracted as firms jockey for individual positions" (Porter 1990, p. 653). When this kind of nonproductive behavior prevails, more industry involvement and autonomy are not necessarily the better option. "In Japan, MITI has frequently applied significant pressure on firms to set basic standards, pushing them to move on to the next stage in the innovation cycle" (Porter 1990, p. 653).[33] In many countries, important industry standards are set by nongovernmental organizations, such as Germany's Deutscher Normenausschuss (responsible for Deutsche Industrie Norm, or DIN), the United States' Underwriters Laboratory, and Japan's Japanese Industrial Standards. Internationally, the ISO (International Standards Organization) has promulgated various standards for industries and business processes.

These experiences suggest that China needs to enhance its standard setting to generate more demand for innovation. Doing so might involve, for example, (1) focusing exclusively on product improvement (such as in performance, fuel efficiency, and environmental impact) and resisting the tendency to use standard setting to protect or help domestic or local industry; (2) taking EU and U.S. standards as a technical starting point while looking for ways to advance product performance; (3) involving industry leaders more in standard setting but ensuring that this is done in a productive way; and (4) changing the government's role from sole standard setter to time-sensitive driver of industrial consensus.

Conclusions

Starting in the mid-1980s, China began departing from a government-led national innovation system in an attempt to make RDI public service units (PSUs) more responsive to the needs of economic development. In more recent years, the role of business enterprises in innovation has been substantially strengthened. Nevertheless, an enterprise-led and market-based national innovation system has not yet been established, as the microinstitutional foundation of such a system—value-maximizing firms and well-functioning markets—remains incomplete. In particular, the role of the private sector in R&D activities is very limited. The bulk of China's R&D activities still has to be performed by LMEs owned and controlled by the state despite the weak incentives to innovate faced by such enterprises. The underdevelopment of market institutions exemplified by distortions in pricing, weak enforcement of regulations, and barriers to free and fair competition also tend to discourage even value-maximizing firms from investing in innovation. In addition, much remains to be done to raise the demand for innovation.

The current situation suggests that a plan to create and strengthen incentives for innovation should have the following key components:

- Promote continuous development of the private sector.
- Further reform SOE governance by focusing on board governance.
- Further reduce the scope of state ownership through means such as dividend collection and secondary share offerings.

- Implement planned reforms in the areas of energy and natural resources pricing; enforce laws and regulations on environmental protection, labor rights protection, and product quality and antimonopoly; and remove barriers to entry, exit, and the free transfer of corporate control through merger and acquisition.
- Improve supply-side incentives such as fiscal incentives to encourage pooled R&D effort locally and globally.
- Improve the use of demand-side instruments, such as government procurement and standard setting, to raise the demand for innovation but with effective protections against rent seeking and protectionism.

CHAPTER 3

Building the Capacity of Private Enterprises

In China's existing national innovation system, state-owned LMEs and RDIs are the main sources of innovation activities. In the future, however, China's success in technological catch-up is likely to rely more on the capacity of its private sector. Although government played more of an active role in promoting innovation in East Asia than it did in Western advanced market economies, none of the East Asian successes—including Japan; the Republic of Korea; and Taiwan, China—can be attributed to a reliance on SOEs. On the contrary, government intervention in those cases was successful only because of collective action by private enterprises that were highly motivated to extend their technological frontiers. Cases in point are the Japanese automobile industry and the SME-led success in upgrading the industrial structure of Taiwan, China (Hou and Gee 1993; Odagiri and Goto 1993). From that perspective, and for the reasons outlined in chapter 2, China must invest in the capacity of its emerging private sector if it is to make the progress it needs both for the creation of technology and for its adaptation and adoption.

China's Emerging Private Enterprises

China's private sector is young and still emerging. However, the magnitude and speed of private business creation since the mid-1990s are remarkable. As of 2006, China had 4.95 million domestic private enterprises, 1.8 times more than in 2000 (Liu and Xu 2007, p. 3).[34] Most of them are small and operate in the service sectors. However, 150,000 of them are industrial enterprises with annual sales revenue of more than Y 5 million (the "cut-off scale").[35] From 1998 to 2006, the number of enterprises in this group rose by a factor of 13 while their total output jumped by a factor of 29 (figure 3.1). The similar trajectories of the three indicators of figure 3.1, i.e., the number of firms above the cut-off scale, their total output, and their share in total industrial output suggests a high correlation between, on the one hand, growth of output of the private sector and the rise of its share in the national economy and, on the other, business creation and growth.

The rapid creation of private businesses has resulted in a private sector dominated by young and small firms run by inexperienced owners. The size of private industrial enterprises, measured by value added per enterprise, is significantly smaller than the sector average in 36 of the 39 industrial sectors reported in China's official statistics (figure 3.2).[36] Data from 2005 on the occupation of business owners before they started their businesses suggest the relatively low level of managerial experience in the private sector (figure 3.3).[37] Former managerial staff and self-employed businesspersons accounted for 42 percent of the total; other owners did

Figure 3.1   The Takeoff of China's Private Sector, 1998–2006

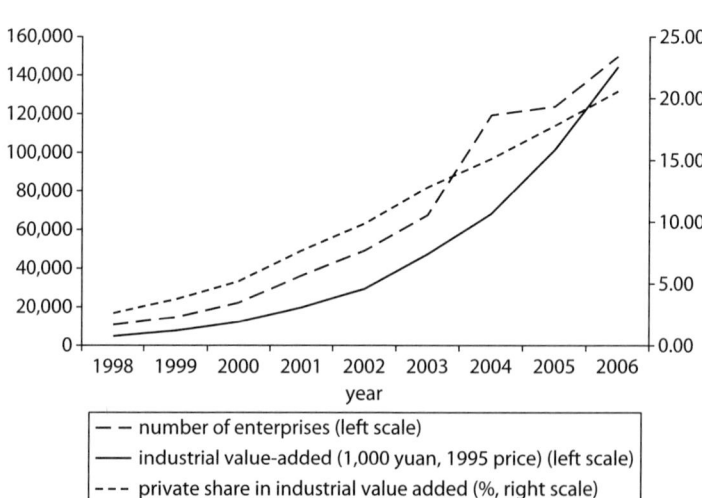

Source: NBS, various years.

Note: Private industrial enterprises with sales revenue of more than Y 5 million.

Figure 3.2 Size of Private Industrial Enterprises Relative to Their Competitors and Their Share in Total Output, by Sector, 2006

120
100
80
60
40
20
0
percent
2 37 1 19 26 34 23 32 21 39 22 17 29 24 35 12 28 3 14 6
sector number
relative size
market share

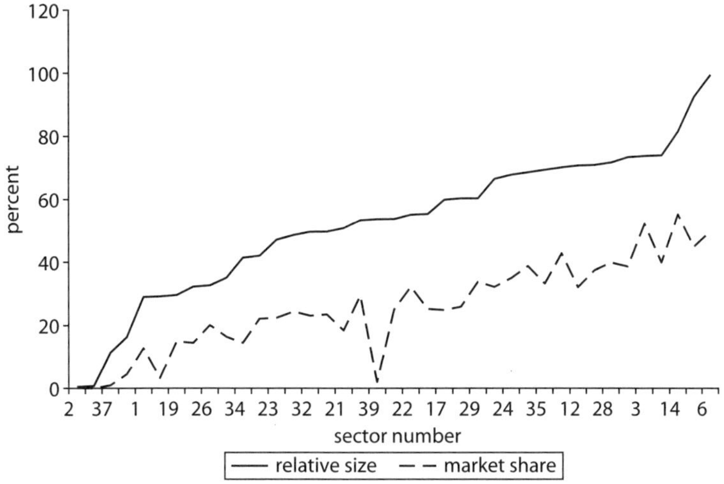

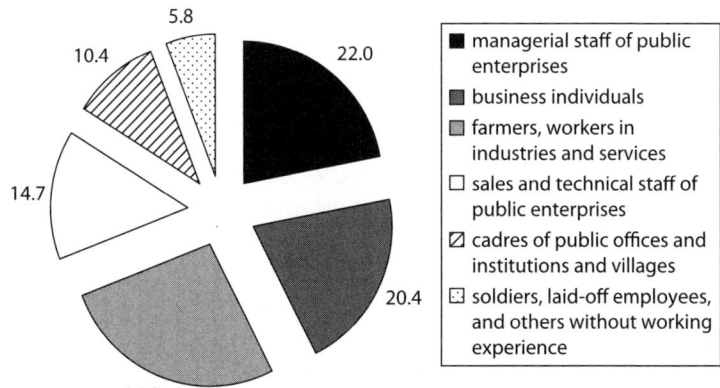

Source: NBS 2007, table 14-10.

Note: Chinese official statistics define 39 industrial sectors, which appear in the *China Statistics Yearbook* in a fixed order, along with the sector names corresponding to the sector numbers in this figure. Relative size is defined as the average size of private enterprises measured in value-added per enterprise as a percentage of the average size of all enterprises in the sector. Market share is the total value added of private enterprises in the sector as a percentage of the sector total.

Figure 3.3 The Occupations of Chinese Private Business Owners before They Started Their Businesses, 2005

(percent)

5.8
10.4
22.0
14.7
20.4
26.8
managerial staff of public enterprises
business individuals
farmers, workers in industries and services
sales and technical staff of public enterprises
cadres of public offices and institutions and villages
soldiers, laid-off employees, and others without working experience

Source: ACFIC 2007, p. 234.

not appear to have much previous business management experience. In particular, the largest group of owners (27 percent) consists of those who were formerly rural and urban workers.

As China continues with industrialization and urbanization, its private sector is likely to continue for some years to be populated by young, small firms run by inexperienced owners. This trend highlights the strategic significance of capacity building of the private sector in China. To the extent that private firms constitute a major part of Chinese industry, closing the distance between Chinese industry as a whole and international technological frontiers depends on the absorptive capacity of the large number of young and small private firms. Anecdotal evidence and various studies show that Chinese enterprises, especially private SMEs operating in traditional sectors and less developed regions, suffer severely from a low capacity to create or absorb technology. They are facing serious constraints in human resources, technology acquisition, and access to various innovation services. The effective implementation of the enterprise-led strategy for innovation must overcome those constraints. Before turning to these specific issues, it will be useful to have a closer look at the results of a survey on the innovation activities undertaken by Chinese SMEs.

Innovation Activities of Chinese Private SMEs: A Close-Up

The Chinese SME Innovation (CSMEI) Survey was carried out for this study in Chongqing municipality and Zhejiang province in late 2006 and early 2007. It was designed by the World Bank study team in collaboration with MOST and was executed by two organizations, the Chongqing Productivity Center and Zhejiang College of Industry and Commerce. Those two organizations selected the 491 SMEs that, when asked, expressed a willingness to participate (244 in Chongqing and 247 in Zhejiang); distributed the questionnaire to them; and received 367 valid responses (202 in Chongqing and 165 in Zhejiang). Responses were provided by the managerial staff of the firms—in most cases, general managers or their assistants. The characteristics of the 367 respondents (table 3.1) can be summarized as follows:

- Most of the firms were established after 1992; the firms established after 2003 were more numerous in Chongqing than in Zhejiang.
- More than four-fifths were domestic private firms.
- Chongqing firms tended to be more technology based, with higher R&D spending, while firms in Zhejiang were more concentrated in

Table 3.1   Selected Characteristics of SMEs in Chongqing and Zhejiang that Responded to the World Bank CSMEI Survey
(*percent except as noted*)

Characteristic	Chongqing	Zhejiang
Number of respondents with valid responses	202	165
Established after 1992	86.6	80.0
Established after 2003	23.3	3.0
Private firm	82.2	87.9
Operates in information technology, pharmaceuticals, new materials, new energy, bio-industry, or environmental technology industries	48.0	5.0
Operates in machinery and chemical industries	43.1	24.3
Operates in food and beverage, home appliance, textile and garment, grocery, and other light industries and agribusiness	8.9	52.5
Located in an industrial park for high-tech industry	30.2	23.6
Self-identified as a "high-tech" firm	67.3	29.7
Self-identified as being in "traditional manufacturing industries without much technology content"	18.3	55.8
R&D spending less than 1% of sales in 2005	10.4	27.8
R&D spending more than 5% of sales in 2005	56.4	33.3
Fewer than 50 employees in 2005	40.1	10.2
More than 500 employees in 2005	18.3	37.5
Sales revenue of more than Y 100 million in 2005	17.5	37.3
Profit of 1–20% of sales in 2005	69.1	70.9

Source: World Bank CSMEI Survey.
Note: For each question, the total number of valid responses varied slightly.

traditional manufacturing, a distribution that allows a balanced exami-
nation of the two kinds of SMEs.
- Firms in Chongqing have lower employment and sales than those in Zhejiang, but the two groups appear to be equally profitable.

What Drives SMEs to Carry Out Innovation Activities?
Respondents were asked to indicate the importance (on a scale of 0–5) of various goals for innovation activities they have carried out over the past three years (table 3.2). The results suggest that the desire for new markets and new products is the primary motivation for R&D in both Chongqing and Zhejiang. However, some fairly distinct differences also characterize the two groups. First, the improvement of existing products is reported as a more important objective by Zhejiang respondents than by their Chongqing counterparts. Second, to a greater extent in Zhejiang than in Chongqing, innovation is driven by regulation and standards or induced by favorable government policies. Third, cost reduction is a more important reason for innovation in Zhejiang than in Chongqing.

Table 3.2   Importance of Selected Objectives of Innovation Activities of SMEs in Chongqing and Zhejiang

			Distribution of valid responses (percent)		
	Region	Number of valid responses	Not an objective or not important	Somewhat or fairly important	Very or extremely important
Develop new markets	Chongqing	200	3.8	26.3	70.0
and products	Zhejiang	162	1.8	27.6	70.6
Improve existing	Chongqing	200	6.2	41.7	52.1
products	Zhejiang	164	3.5	34.7	61.7
Comply with regula-	Chongqing	200	11.1	50.3	38.7
tions and standards	Zhejiang	159	2.5	41.3	56.3
Qualify for preferential	Chongqing	199	11.6	50.8	37.7
policies	Zhejiang	160	8.8	44.4	46.9
Reduce costs	Chongqing	194	13.2	59.3	27.6
	Zhejiang	158	7.2	45.9	46.9

Source: World Bank CSMEI Survey.

Note: The question was as follows: Has each of the following been an objective of your firm's innovation activities over the past three years? If yes, how important has it been? 0 = Not an objective; 1 = Not important; 2 = Somewhat important; 3 = Fairly important; 4 = Very important; 5 = Extremely important. Eleven objectives were presented. Values for some objectives in the table are the arithmetic average of responses for more than one objective.

How Are Innovative Activities of SMEs Usually Carried Out?

The respondents were also asked to rank the importance of some strategies for carrying out innovation activities (table 3.3). The results indicate that firms in both Chongqing and Zhejiang regard R&D team building as the most important strategy. However, Chongqing firms tend to rely much less on external resources than do their Zhejiang counterparts. For example, nearly half of Chongqing respondents reported that outsourcing to RDIs and HEIs was not an important strategy, while only about one-fourth of Zhejiang respondents expressed the same view. A notably higher percentage of Zhejiang firms also attached great importance to collaboration with their suppliers and clients.

The sharpest contrasts were found in three areas. First, nearly half of Chongqing firms did not think collaboration with their competitors was important for their innovation activities, whereas 70 percent of Zhejiang firms marked it to be anywhere from fairly important to extremely important. Second, taking over other firms was regarded as unimportant by about 70 percent of Chongqing firms, but a nearly equal proportion of Zhejiang firms treated it seriously. Third, the purchase of more-advanced equipment was regarded as at least somewhat important

Table 3.3   Importance of Selected Innovation Strategies of SMEs in Chongqing and Zhejiang

			Distribution of valid responses (percent)		
	Location	Number of valid responses	Not a way or not important	Somewhat or fairly important	Very or extremely important
Building own	Chongqing	200	7.0	44.5	48.5
R&D team	Zhejiang	159	4.4	44.7	50.9
Outsourcing to	Chongqing	196	51.5	37.8	10.7
RDIs or HEIs	Zhejiang	154	24.7	40.3	35.1
Collaborating	Chongqing	195	20.5	50.8	28.7
with suppliers	Zhejiang	162	6.8	41.4	51.9
Collaborating	Chongqing	196	12.8	44.4	42.9
with clients	Zhejiang	162	8.0	34.0	58.0
Collaborating with	Chongqing	194	48.5	45.4	6.2
competitors	Zhejiang	153	30.7	34.0	35.3
Purchasing patents	Chongqing	194	50.5	35.6	13.9
and licenses	Zhejiang	155	16.1	42.6	41.3
Purchasing more-	Chongqing	199	22.6	46.2	31.2
advanced equipment	Zhejiang	161	3.7	37.9	58.4
Taking over	Chongqing	188	72.3	22.3	5.3
other firms	Zhejiang	152	34.9	29.6	35.5

Source: World Bank CSMEI Survey.
Note: The question was as follows: Has each of the following been a way for your firm to carry out innovation activities over the past three years? If yes, how important has it been? 0 = Not a way; 1 = Not important; 2 = Somewhat important; 3 = Fairly important; 4 = Very important; 5 = Extremely important.

by almost all Zhejiang firms, but it was rated as unimportant by more than one-fifth of Chongqing firms. These differences may reflect the difference between the two groups in their dependence on high technology. But they might also be an indication that R&D at Chongqing firms could benefit from a more open stance toward external opportunities and resources.

How Do SMEs Tap into the Knowledge Pools of RDIs and HEIs?

Chongqing firms indicated that they used RDIs and HEIs mostly for technical consulting and joint research (table 3.4). Technical consulting is also a major mode of cooperation with RDIs and HEIs for Zhejiang firms, but they use training more than joint research. In both regions, more than 15 percent of respondents did not have any type of cooperation with RDIs and HEIs.

Table 3.4   Modes of Cooperation with RDIs and HEIs by SMEs in Chongqing and Zhejiang

	Number of valid respondents		Percentage of all valid respondents	
Mode	Chongqing	Zhejiang	Chongqing	Zhejiang
Technical consulting	110	68	54.5	42.0
Joint research	113	40	55.9	24.7
Management consulting	45	38	22.3	23.5
Training	53	61	26.2	37.7
Other	7	3	3.5	1.9
No such cooperation	31	30	15.3	18.5
Memo: Total number of respondents	202	162	100.0	100.0

Source: World Bank CSMEI Survey.
Note: The question was as follows: Which of the following kinds of cooperation does your firm have with RDIs and HEIs (multiple choices are allowed): technical consulting, joint research, management consulting, training, other, no such cooperation?

What Are the Key Difficulties Facing SMEs in Their Innovation Activities?

To gain a sense of the difficulties surrounding innovation by SMEs, respondents were asked to indicate which of the following situations were close to their own experience over the past three years:

- Had an idea for an innovation but did not implement it because of
  - lack of funding
  - lack of talent
- Had an idea for an innovation and conducted R&D but was not successful because of
  - technical reasons
  - economic reasons
- R&D was successful but failed to be commercialized because of
  - lack of market demand
  - lack of capital
  - policy restrictions

Respondents were allowed to mark more than one item. A shortage of talent seems to be the strongest common characteristic of firms in Chongqing and Zhejiang; in both groups, one-third to one-half of respondents reported the problem (table 3.5). In addition, both groups reported that policy restrictions were not a significant difficulty in commercializing R&D results. In other respects, however, the two groups differed

Table 3.5   Causes of Unsuccessful Innovation Activities of SMEs in Chongqing and Zhejiang

Situation	Cause	Number of respondents		Percentage of all respondents	
		Chongqing	Zhejiang	Chongqing	Zhejiang
Had idea but did not conduct R&D	Lack of funding	104	33	52.0	29.5
	Lack of skilled workers	68	50	34.0	44.6
Conducted R&D but was unsuccessful	Technical reasons	51	56	25.5	50.0
	Economic reasons	68	14	34.0	12.5
R&D successful but was not commercialized	Lack of market demand	64	29	32.0	25.9
	Lack of capital	70	8	35.0	7.1
	Policy restrictions	17	10	8.5	8.9
Total number of respondents	n.a.	200	112	100.0	100.0

Source: World Bank CSMEI Survey.
Note: n.a. = not applicable.

significantly. First, capital constraints seem to have been much worse in Chongqing than in Zhejiang. Second, Chongqing firms seem to have experienced more difficulty in terms of economic factors and lack of demand than did their Zhejiang counterparts. However, technical problems were much more frequent in Zhejiang than in Chongqing. The general picture seems to be that both groups of firms are short of R&D talent; Chongqing firms are constrained more than Zhejiang firms by economic and financing problems; and Zhejiang firms have a lower technical capacity for R&D.

Effectively Managing Human Resources for Innovation

As seen in the preceding section, a shortage of talent is perceived by both high-tech and conventional industrial SMEs in the World Bank CSMEI Survey as a major difficulty for innovation activity. Indeed, adequately educated and skilled workers are critical to innovation, in terms not only of technology creation but also of adaptation and adoption. What is the nature of the problem, and what actions are likely to resolve it?

Issues

The finding that a shortage of talent is a primary obstacle to innovation by private Chinese firms appears in other recent empirical studies. For example, in an unpublished survey conducted in 40 cities in China by the

NBS in 2006 (the NBS survey), "shortage of technological personnel" was chosen by respondents as the second-highest barrier to innovation, just below "shortage of funding." In a survey of private enterprises conducted in 2006 by the Chinese Academy of Social Sciences (the CASS survey), 66 percent of the 1,594 sample firms believed that a shortage of talent hindered technological innovation in their firms, while only 50 percent put "lack of capital" on the list (Liu and Xu 2007, p. 21).

However, this perception, as commonly held as it is, needs to be interpreted with caution. It is useful to go further to explore exactly what the causes of the perceived talent shortage are: Is it due to market constraints in that there are just too few skilled workers or barriers between them and their potential employers? Or is it essentially a result of inadequate internal management of human resources (HR)?

In the World Bank CSMEI Survey, respondents were asked to indicate the seriousness of six HR issues in the context of innovation activities. The most important reported issue is the concern that technical secrets will be taken away by R&D workers who resign from the firm, a concern that seems to be shared by respondents from both Chongqing and Zhejiang (table 3.6). Another shared issue is the difficulty of retaining skilled R&D workers, a problem that might be closely linked with the first one. That is, if employers are seriously concerned about the risk of losing technical secrets when employees leave, the firms' precautionary measures—for example, granting only limited trust to R&D workers—may well reduce their ability to retain those workers.

Compensation is also seen as a challenge by firms in both locations, albeit to a lesser extent in Chongqing. However, the availability of skilled R&D workers and access to them are perceived as serious problems by Zhejiang firms but much less so by Chongqing firms; in particular, as much as 50 percent of Chongqing respondents did not think those issues posed any problem to their firms. Chongqing firms also appear to be significantly more confident of their HR management skills than are their Zhejiang counterparts. Overall, Zhejiang firms seem to face severe market and management constraints, whereas management constraints appear to be more severe for Chongqing firms than market constraints despite the higher confidence of those firms in their HR management experience.

Clearly, the issue is not a simple one of talent shortage in the usual quantitative sense, in which firms are unable to hire workers with the needed skills at the salary they are willing to offer.[38] Both market constraints and management constraints play a role in causing the perceived talent shortage. However, firms in different regions and sectors appear to

Table 3.6   Issues of HR Management in SMEs in Chongqing and Zhejiang

			Distribution of valid responses (percent)			
Issue	Region	Number of valid responses	Not applicable or not a problem	A small problem	Important or thorny problem but can be solved	A thorny problem and hard to solve
We lack	Chongqing	198	31.8	21.7	41.9	4.5
experience in HR management	Zhejiang	156	11.5	5.8	76.3	6.4
Regarding the skilled R&D workers we want						
We are not sure	Chongqing	185	49.2	20.5	28.1	2.2
whether they are available	Zhejiang	152	10.5	15.1	64.5	9.9
We are sure	Chongqing	190	41.6	17.9	33.7	6.8
they are available, but we cannot locate them	Zhejiang	153	12.4	15.0	57.5	15.0
We can locate	Chongqing	192	27.1	16.7	43.8	12.5
them, but we cannot meet their demands for compensation and welfare	Zhejiang	150	12.7	15.3	56.7	15.3
It's difficult for us	Chongqing	189	21.7	15.9	51.3	11.1
to retain skilled R&D workers	Zhejiang	151	15.2	9.3	58.9	16.6
We cannot	Chongqing	185	19.5	13.0	37.8	29.7
prevent R&D workers from taking our technical secrets when they resign	Zhejiang	152	14.5	6.6	54.6	24.3

Source: World Bank CSEMI Survey.
Note: Respondents were asked to assign a value to the issues on a 0–5 scale: 0 = Not applicable; 1 = Not a problem; 2 = A small problem; 3 = An important problem; 4 = A thorny problem but can be solved; 5 = A thorny problem and hard to solve.

face different situations in terms of availability of, and access to, R&D skills, while HR management stands out as a common constraint to all firms.

Ownership and governance seem to have also played a role in the perceived shortage of R&D workers. In the World Bank CSMEI Survey, respondents identified themselves in terms of the following category of ownership and governance: sole proprietorship, partnership, family-controlled business, business controlled by multiple shareholders, state-owned enterprises, collectively owned enterprises, and FDI enterprises. Response by ownership category on the seriousness of the two most significant challenges in HR management—the difficulties of keeping technical secrets and of retaining skilled R&D workers—shows a distinct difference between domestic private firms, on the one hand, and public and FDI firms, on the other (table 3.7). In particular, the significantly lower values assigned by FDI firms to the two difficulties suggest the extent to which the difficulties stem from internal governance and management rather than external market conditions.

Evidence from the CASS survey also suggests the rudimentary nature of HR management in Chinese private enterprise, but from a different perspective: that of firms' commitment to employee training. Average spending on training for the sample firms in 2004–06 was in the range of 1.5–4.6 percent of the wage bill (table 3.8). For a firm with an average wage bill of Y 12,000 per worker, that ratio would imply an annual training budget of merely Y 180–Y 552 ($24–$75) per worker. Training spending is related to employee retention (figure 3.4).

Table 3.7   Difficulties in Managing Skilled R&D Workers, by Firm Ownership and Governance Categories

Ownership and governance	Number of valid respondents	Average value of valid responses on difficulties in keeping technical secrets	Average value of valid responses on difficulties in retaining skilled R&D workers
Sole proprietorship or partnership	99	3.0	3.4
Family controlled firm	33	3.2	3.1
Company controlled by multiple shareholders	131	3.0	3.3
State and collective firms	39	2.9	3.2
FDI firms	22	2.4	2.5

Source: World Bank CSMEI Survey.
Note: Thirty-two respondents chose more than one category of ownership and governance. They have been assigned to one of the categories according to the combination of their choices (for example, a firm describing itself as a partnership and a family business is treated as a family business; a firm identifying itself as a state enterprise with multiple shareholders is treated as a state enterprise). Values assigned in the responses: 0 = Not applicable; 1 = Not a problem; 2 = A small problem; 3 = An important problem; 4 = A thorny problem but can be solved; 5 = A thorny problem and hard to solve.

Table 3.8   Average Training Expenses and Staff Turnover in Chinese Private Enterprises, 2004–06, by Sector

Sector	Number of firms in sample	Training expenses as percentage of wage bill	New hiring as percentage of total employment	Resignations as percentage of total employment
Agriculture, forestry, animal husbandry, and fishing	48	2.36	23.45	14.62
Transport	20	1.51	9.51	6.44
Foods and beverages	41	4.36	17.35	14.95
Construction and real estate	75	2.58	11.92	6.83
Paper and printing	27	3.09	16.00	11.13
Chemicals	27	3.55	13.35	9.16
Vehicles and motors	40	3.75	23.16	14.39
Hardware	37	3.70	18.89	12.36
Machinery and Electronics	116	4.62	18.67	9.92
Metal smelting	32	2.06	16.71	8.56
Light and textile	73	2.42	17.85	11.66
Plastics and rubber	24	2.80	17.22	10.16
Construction materials	30	4.59	14.49	9.52
Trade and commerce	53	3.07	15.05	10.88
Catering services	47	4.28	20.24	16.18
Pharmaceuticals and biotechnology	27	3.01	15.74	10.42

Source: Liu and Xu 2007, p. 100.

Figure 3.4   Training Expenses and Employee Resignations in Chinese Private Firms, 2004–06

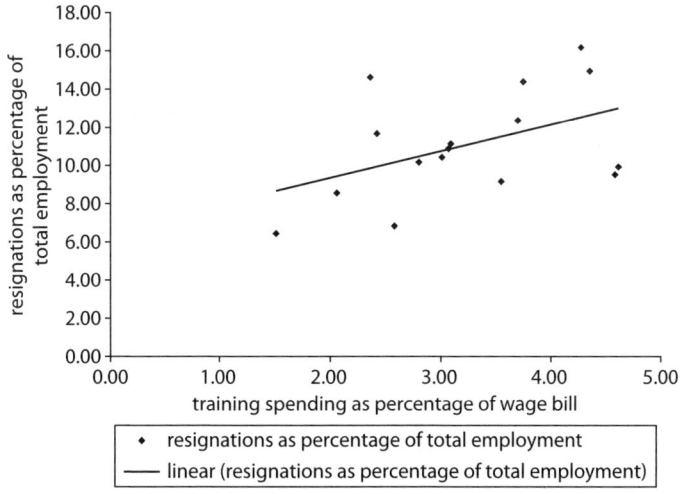

Source: Table 3.8, this volume.

Recommendations

The issues detailed earlier suggest that improvement can be made by modernizing HR management, protecting the firm's technical secrets with legal means, better adapting to labor market conditions, and encouraging employee training.

Modernizing HR management. Great potential seems to exist for private SMEs to modernize their HR management. An overwhelming majority of Chinese private SMEs started as family businesses, which are often characterized by an informal, opaque, and relation-based style of HR management. For a firm to grow into a large corporation and to be innovative, it must modernize its HR management to institute a rules-based system. Some firms have made the transition, but many others are either still struggling or have failed. The reason cited most often for the difficulty in keeping skilled workers is a management style and corporate culture based on kinship, which many private entrepreneurs are unwilling or unable to change even when their firms are desperately short of skilled employees.

One potential reason for the lack of progress in modernizing HR management could be that many entrepreneurs attach too much value to short-term financial gain and too little to employee loyalty and morale. For example, the 2006 survey conducted by ACFIC, which covered 3,837 private enterprises, found that, although the situation has been improving, only a small fraction of the firms have joined national social insurance programs, and only a small fraction of their employees have been covered (table 3.9). In addition, about 27 percent of the sample enterprises did not sign employment contracts with their employees (ACFIC 2007, p. 246).

Table 3.9   Social Insurance in Chinese Private Enterprises, 2006

Social insurance plan	Percentage of sample enterprises that have joined the plan	Percentage of employees covered by the plan in enterprises that responded to the question	Enterprise expenditure per covered worker for the plan (yuan)
Pension	43.9	29.2	2,921
Health	36.9	26.3	2,435
Unemployment	22.2	18.2	1,137
Work injury	24.4	10.7	838
Maternity	13.9	n.a.	n.a.

Source: ACFIC 2007, p. 246.
Note: n.a. = not available.

For many owners and managers of private SMEs, modernizing HR management will involve a change of mindset. Given the different nature of innovation activities from traditional production activities, more emphasis must be put on the loyalty, motivation, and morale of employees as the firm's business becomes more knowledge based, and the emphasis on performance evaluation focusing on measurable results must be lessened. In this regard, the experience of the Indian information technology (IT) services firm HCL Technologies, which has a policy of "employees first, customers second," is enlightening (box 3.1).

Box 3.1

"Employee First, Customer Second": A Soft Slogan with Hard Value at HCL Technologies

Founded in India in 1974 as a computer hardware manufacturer, HCL Technologies is a global IT services company and claims to be India's fourth-largest firm specializing in outsourced software, remote infrastructure management, and other business process outsourcing. As of June 30, 2007, HCL Technologies had revenues of $1.7 billion and 52,000 employees in 17 countries.[a]

The "Employee First, Customer Second" (EFCS) corporate strategy was first introduced to HCL employees in July 2005 by Vineet Nayar, the new president of HCL Technologies, as part of a five-year plan to rejuvenate the company. At the time, HCL was faced with grave challenges in maintaining its growth and especially in retaining talent and sustaining employee morale in the midst of India's increasingly competitive IT industry.

The seemingly soft slogan of EFCS is not about rolling out the red carpet to please employees. It is based on the hard value that effective employee interactions with customers are critical for a service business like HCL. The initiative tends to make result-oriented investment in employees by creating the mechanism and culture in which employee empowerment and development are highly valued and employees become willing and able to practice the corporate value and unleash their potential in reaching customer satisfaction with unique, innovative, and sophisticated services. Here are some elements of the EFCS strategy:

- "360-degree feedback": An annual performance review mechanism in which all employees get feedback on their performance from their managers, peers, and subordinates. This is intended to increase the accountability of managers to their subordinates.

(continued)

Box 3.1 *(Continued)*

- The corporate intranet: A platform for the company to deliver information to all HCL employees as well as to receive employee comments and resolve complaints. The results of the 360-degree feedback for the top 20 managers, including the president, are posted there. Through the Smart Service Desk, an employee can enter complaints on issues ranging from the size of bonuses to the quality of food in the company cafeteria, and only that employee can determine if the case has been satisfactorily resolved. Employees can also post questions for the HCL president, who answers about 50 each week.
- "Trust pay": HCL has instituted so-called trust pay for its engineers as a replacement for the performance-based bonuses that are common in the industry. Under trust pay, engineers may receive their full bonuses for team projects even if the project has not yet met its goals.

In the three quarters after the initiation of the EFCS program, HCL's troubling attrition rate (once among the highest in the industry) significantly dropped, from 20.4 percent to 17.2 percent,[b] and sales rose substantially.

Sources: Hill, Khanna, and Stecker 2007; http://en.wikipedia.org/wiki/HCL_Technologies; http://www.businessweek.com/magazine/content/07_47/b4059064.htm?campaign_id=rss_null; and http://www.globalservicesmedia.com/Content/general200705211097.asp.
a. http://en.wikipedia.org/wiki/HCL_Technologies.
b. McGregor 2007.

The enactment of the Labor Contract Law in 2008 provides an opportunity for private SME owners to catch up with best practice in HR management by observing the new legal requirements for labor rights protection. Each level of government (through their SME departments) and many private sector organizations (ACFIC and industrial associations, for example) are in a good position to provide assistance and guidance. As discussed in the next section, the new law also points the way for firms to better protect their technical secrets.

Making full use of the Labor Contract Law. The Labor Contract Law provides legal instruments with which employers can protect themselves from the risk of losing technical secrets to departing employees. Article 22 of the law provides that employers may establish a confidentiality agreement (*baomi xieyi*) that imposes a legal duty on employees to protect certain commercial and technical secrets for the benefit of the employer. Article 23 further states that the employer may establish a competition restriction (*jingye xianzhi*), a period after the employee leaves during which the employee is obliged not to work for competitors of the employer, and for which the employer will pay the employee. Private SMEs may benefit from making full use of confidentiality agreements and competition restrictions, and the government is well advised to ensure that these rules are adequately enforced. Removing employers' worries over the loss of technical secrets to departing employees could help speed the process of modernizing HR management.

Adapting better to labor market conditions. Actions could also be taken by the private sector and the government to overcome external market constraints. The labor market provides the channel for SMEs to tap into the regional and national pools of R&D talent. However, they must adapt to market conditions if they are to get the skilled workers they need. Such adaptation includes, for example, using employee recruitment ("headhunting") services; establishing regular exchange channels with HEIs, training institutions, and employment centers; and advertising positions on the Web. Adequate provision of related services is equally important. For example, one measure by which the government could help SMEs better manage and tap into the limited pool of skills is to establish institutions such as SME skills development centers at the local level, which could be publicly owned but privately operated (box 3.2).

Encouraging training and lifelong learning. The government could also help SMEs by strengthening policies supporting training and vocational education. For example, the government currently provides tax incentives to encourage enterprises to provide employee training, but the maximum training expenditure eligible for a tax benefit is 2.5 percent of the firm's wage bill.[39] That limit appears to be low given the educational background of the segment of the labor force that SMEs are facing and the rapid pace of business creation and expansion. Even in the period 2004–06, most firms covered in the CASS survey were already spending more than 2.5 percent of their wage bills on training (table 3.8). Although expenses exceeding the 2.5 percent cap can be carried over for a tax benefit in the following year, that allowance is not meaningful if most firms exceed the 2.5 percent cap every year. The high turnover shown in the CASS survey (table 3.8) suggests a high externality of employer spending on employee training—that is, the benefit of the training of an employee accrues to other employers and the society while the cost is born by the employer

Box 3.2

A Three-Part Mission for an SME Skills Development Center

The first mission of an SME skills development center is to provide management and technical training to SME owners and managers—especially the aspects related to innovation, such as entrepreneurship, human resources management, innovation risk management, fund-raising, venture capital, intellectual property rights, technology commercialization, and so on. The curricula should be market driven and developed by business training professionals in collaboration with entrepreneurs and managers. A potentially useful model is the Penang Skills Development Center (PSDC), in Malaysia. The PSDC "operates as a non-profit organization . . . Participating companies pool their resources together to help plan, design, and conduct an extensive range of training programs directly relevant to immediate and forecasted needs. This enables PSDC to offer the most cost-effective training for the industry and at the same time bridge the gap between skills taught in public institutions and skills acquired on the job."[a]

The second mission of an SME skills development center is to provide information on supply and demand conditions for particular skills and on the pay premium for various job categories. The information could be developed and provided through close collaboration with schools, training institutions, and the labor market. This mission focuses on information, not actual headhunting, which can be undertaken in the market.

The third mission is to collect and disseminate, through the center's training program, success stories about skills management and the promotion of an innovation culture within the firm, especially stories from inside China.

a. http://www.logos-net.net/ilo/150_base/en/init/mal_5.htm.

who spends on the training—and, hence, a high return to government investment. A review of the merit of the 2.5 percent ceiling is advisable. For vocational education, the government should consider recommendations provided by another World Bank study (box 3.3).

Facilitating the Collaboration of SMEs with RDIs and HEIs

Given their limited size, SMEs need to access external sources of information, knowledge, know-how, and technology to strengthen and scale up their capacity to create and absorb technology. Collaboration with RDIs

Box 3.3

Building a System of Lifelong Learning in China

China has made impressive strides in expanding access to education at all levels, improving adult literacy, and providing training and retraining to millions of rural migrants and urban workers. However, given today's highly competitive global economic environment and the increasing demand for education and training, China needs to develop a more effective system of lifelong learning to cope with various challenges.

To develop such a system, the government of China needs to take a new role in education—moving from being the main provider of education and training to being the overall architect and facilitator of a more complex system featuring multiple pathways and multiple providers. In shifting into this new role, the government needs to build close partnerships with the private sector and other stakeholders to accomplish the following goals:

- Set up the rules of the system to ensure the quality, relevance, efficiency, and equity of education and training. To that end, put in place systems of accreditation and certification as well as a comprehensive qualification framework and standards. The increasing regional and rural-urban gaps have to be addressed seriously.
- Provide accurate and timely information on changing market demands; employment opportunities; and the quality, performance, and offerings of education and learning providers (including international providers).
- Provide sufficient funding for the increasing demand for education and training. As government focuses more on compulsory education, the private sector can play a bigger role in high-end education and training.
- Harness the potential of distance education. Although distance education is expanding rapidly, more attention needs to be paid to improving its quality and recognizing its value to the society and labor market.

Source: Dahlman, Zeng, and Wang 2007.

and HEIs is an important way for SMEs to tap into pools of knowledge and technology. How well is the current level of collaboration serving SMEs, and to what extent is the work of the RDIs and HEIs directed at them?

Issues

Looking at the issue from the demand side, more than three-fourths of respondents had a service contract with either an RDI or an HEI over the preceding three years, according to the World Bank CSMEI Survey. The survey attempted to gain a sense from the firms of how well the institutions executed the contracts and how well they provided the services requested. Most firms in both Chongqing and Zhejiang rated the execution of the contract and the quality of service as "just acceptable." However, Chongqing firms appear to have been more satisfied with the execution of contract and service quality (30–40 percent marked them as "excellent" or "very good") than were their Zhejiang counterparts (around 25 percent chose "excellent" or "very good"), suggesting a lower degree of satisfaction on the part of firms operating in traditional industries (table 3.10).

Looking at the issue from the supply side, the available data point to a possible unrealized potential in this area, at least in the case of RDIs. With the reform of PSUs in the direction of "pushing to the market" (World Bank 2005), Chinese RDIs have strengthened their cooperation with business enterprises over the past two decades. However, the overall situation remains worrisome: of the 260,242 person-years that RDIs invested in "S&T projects" in 2006, only 5.3 percent was invested in "enterprise contracted" projects and only 2.9 percent in "enterprise collaborated" ones (table 3.11). In terms of financial inputs, the shares of these two kinds of projects are 5.8 percent and 1.9 percent, respectively. The role of enterprises in "R&D projects," a subset of "S&T projects," is even weaker; that suggests a fairly large potential to further encourage

Table 3.10   Experience of SMEs in Chongqing and Zhejiang with Contract Execution and Services of RDIs and HEIs

Item	Contract execution			Service quality		
	Chongqing	Zhejiang	Total	Chongqing	Zhejiang	Total
Number of respondents	202	156	358	201	155	356
Percentage of respondents who had contracts	76.2	74.4	75.4	80.6	74.8	78.1
Experience of firms as a percentage of those with contracts						
Very bad	0.6	4.3	2.2	0.6	1.7	1.1
Bad	11.7	17.2	14.1	10.5	12.1	11.2
Just acceptable	45.5	51.7	48.1	57.4	61.2	59.0
Very good	33.8	18.1	27.0	28.4	14.7	22.7
Excellent	8.4	8.6	8.5	3.1	9.5	5.8

Source: World Bank CSMEI Survey.
Note: The questions were as follows: How would you rate the execution of the contract between your company and HEIs/research institutes? How would you rate the quality of services provided by HEIs/research institutes?

Table 3.11   Role of Enterprises in S&T and R&D Projects of RDIs in China, 2006

RDI project type	Number of Projects	Employment involved (full-time equivalent person-years)	RDIs' expenditure (10,000 yuan)
S&T			
All	64,169	260,242	451,7195
Enterprise-contracted (percent)	9.3	5.3	5.8
Enterprise-collaborated (percent)	4.7	2.9	1.9
R&D (subset of S&T)			
Number	42,262	202,360	365,3731
Enterprise-contracted (percent)	4.3	3.5	3.4
Enterprise-collaborated (percent)	3.3	2.1	1.3

Source: NBS and MOST 2007, pp. 86, 89.
Note: According to MOST officials in communication with the study team, R&D projects are included in S&T projects, and enterprise-contracted projects can overlap with enterprise-collaborated projects. Enterprise-collaborated projects include projects that involve collaboration with enterprises through means other than a contract awarded to a RDI from an enterprise.

knowledge institutions, especially PSU RDIs, to increase their enterprise-driven R&D activities.

Interviews with local government officials and private entrepreneurs by the World Bank's study team in Zhejiang province confirmed findings of some earlier studies (Motohashi 2006) that the industry-research link is partially undermined by the short-term focus and opportunistic behavior of collaborating parties. Many unsuccessful cases of collaboration were attributed to two factors. The first was the business owners' excessive focus on short-term gains and unwillingness to risk failure. The second was the researchers' excessive concentration on the implications of the assignment for income and professional title advancement. Some other factors may have hindered more effective collaboration between SMEs and collaborating institutions. For example, a previous study found that despite recent improvements, the still-existing social stereotype and discrimination against private enterprises (most of them SMEs) make it harder for them to establish formal collaboration mechanisms with RDIs and HEIs, which are mostly public. In addition, the current forms of contractual relationship lack a long-term mechanism for sharing benefits and are vulnerable to dispute (Liu and Xu 2006).

Recommendations

To help SMEs better tap into the knowledge and skill pools of innovation networks, the government could consider initiatives that use measures especially appropriate to SMEs. The following examples draw from relevant international and Chinese experiences.

SME access to information. Many governments operate programs, such as Internet-based services, to improve SME access to information about networking opportunities. One example is Canada's Strategies and Innovation Portal, a Web site launched in 1996 by Industry Canada. It includes a comprehensive inventory of links to innovative HEIs, public laboratories, federal and municipal agencies, and businesses across Canada. The site includes information on business intelligence, financing, human resources, product development, marketing, intellectual property, and research services. Also included are diagnostic tools and hundreds of subjects related to innovation.[40] Another example is the Shanghai R&D Public Service Platform, which enables SMEs and other users to tap into the rich pool of R&D resources scattered across many companies and institutions in Shanghai.[41] Through a telephone conversation or online chat with a staff member of the R&D platform, users can be put in contact with resource suppliers. The resources that can be shared through the platform include literature, data, instruments, testing facilities, measuring facilities, advice from technical professionals, and technology transfer services. As of May 2008, the platform had 130,000 registered users, 30 percent of whom were from outside Shanghai.[42]

Technology "brokering" programs. Innovation brokerages, if properly operated, can play an instrumental role in stimulating less innovative firms to become more innovative as well as promoting the formation of networks and interactive learning among firms and knowledge institutions. A well-known example of such a brokerage is the Norwegian TEFT (Technology Diffusion from Research Institutes to SMEs), which uses "technology attachés" as brokers (box 3.4). These attachés act variously as analysts, brokers, mediators, and coaches as they work in a proactive manner to raise the technology capacity of the firms. The program is able to motivate less innovative SMEs to cooperate more with knowledge institutions through networks and perform joint innovation projects. However, the formation of such networks is a long-term process that requires the active development of mutual interests and trust on the part of both parties (OECD 2004). In addition to a large number of public RDIs, China also has many technology diffusion organizations, such as engineering research centers and productivity centers, which are often PSUs. The government should consider piloting brokerage programs, suitably adapted to

Box 3.4

The TEFT Technology Attachés as Brokers

Started in 1994, TEFT (Technology Diffusion from Research Institutes to SMEs) is a nationwide program that aims to encourage SMEs to become more R&D conscious by developing closer links between them and the five largest polytechnic R&D institutions in Norway. At the same time, the program aims to change the R&D institutions' attitudes toward SMEs and to strengthen their knowledge of the innovation needs of SMEs. The program spans the whole spectrum of Norwegian industry, but it is primarily intended to reach sectors with low or average levels of R&D and companies with 10–100 employees.

The link between SMEs and R&D institutions is the TEFT attaché. Each of the attachés, who are seconded from the R&D institutions, is responsible for a specific geographic area and acts as a broker, organizer, or coach to aid SMEs in the innovation process. The attachés maintain an active program of visits to companies and are normally the companies' first contact with the TEFT program. A technology project begins with an evaluation of the enterprise by the technology attaché for that firm's area. The attachés are in close touch with what is going on in Norwegian technological research institutions, and they put companies in contact with scientists, who carry out the technology project in close cooperation with the company. The TEFT program pays 75 percent of project costs, and the company pays the remaining balance. Financial support from TEFT will usually come to between EUR 4,000 and EUR 13,000.

In these ways, TEFT aims to lower barriers to cooperation between national R&D institutions and SMEs. The evaluation of the TEFT program by Norwegian consultants was generally positive in terms of the increased cooperation between industry and research environments and particularly with respect to the "go-between" role of the technology attachés. Overall, firms supported by the program report significant improvements in products and production technology as well as an increase in R&D intensity and capability. Participation in TEFT has led to the following specific results: improvement in existing products (43 percent of the firms supported by the program from late 1990s to early 2000s), new products to the firm (35 percent), improved production technology (40 percent), increased R&D (41 percent), and increased R&D capacity (59 percent). However, evaluations also pointed out that the TEFT program's emphasis on regional development was greater than its emphasis on innovation and economic growth. Overall, the evaluations point to a good match between the needs of the firms and the supply of the research organizations: 73 percent of the firms report that there is a good link between TEFT and the firms' business plan and 85 percent of the firms report that they collaborated easily with the TEFT researchers.

Sources: Asheim, Isaksen, Nauwelaers, Todtling 2003; EC 2002; OECD 2004; The Scottish Government 2007.

local conditions, which would bring RDIs and other public institutions together with SMEs in a proactive manner.

SME participation in public-private partnerships. In government-financed R&D programs for businesses, the government provides grants for some strategic sectors and invites private sector firms to compete with each other to participate in R&D with public institutions. Involving SMEs in such pubic-private partnerships is essential for the stimulation of technological entrepreneurship, for SMEs to gain access to knowledge sources, and for the linking of science-based innovation networks to less R&D-intensive ones. SMEs overall do not make much use of R&D support, nor do R&D support systems tend to target SMEs, so a preference for small firms in government R&D programs helps reduce the "bias" against SMEs. Drawing on the Dutch experience (box 3.5), the government at both the central and the local levels in China can consider testing this idea.

Box 3.5

The Innovation Voucher for SMEs in the Netherlands

In the Netherlands, SMEs receive innovation vouchers from the government to be spent on research, whether basic or applied, to be supplied by government-run knowledge institutions such as HEIs and technology transfer organizations. Initially, the value of the voucher was EUR 7,500. In 2004 and 2005, 850 vouchers were allocated in three rounds. In 2006, the government made more vouchers available, lowered the value, and requested SMEs to pay at least one-third of the cost of the assignment. Studies have found that the innovation voucher stimulates SMEs to engage in many projects with government knowledge institutions that otherwise would not have been attempted: 80 percent of the vouchers are used for projects that would not have been assigned without such a voucher, 10 percent are used for projects that would have been assigned without them, and 10 percent are not used. There is evidence that some of the assignments in the 80 percent group would actually have been commissioned later in any case, but have been ordered earlier because of the voucher. The studies also found a significant positive effect of the innovation voucher on process improvement. However, the long-term effects of these vouchers on innovation still need to be studied.

Sources: CPB 2007; OECD 2005b.

Personnel mobility programs. Mobility programs encourage enterprises to give internships to graduate engineering students or research scientists at HEIs or RDIs to work in the firm for several months. Likewise, such programs support enterprises in the temporary placement of their engineers and technical personnel at HEIs or RDIs. Many OECD countries have adopted measures to support such temporary placements and industry-funded PhD projects (Huang and others 2005). One example of good practice is the United Kingdom's Business Fellowship program, through which mainstream academics spend part of their time advising companies on technical or research problems.

Some countries provide SMEs with inducements to participate in mobility programs. Denmark has introduced a tax deduction on collaborative R&D. In Belgium, the First-Enterprise program covers up to 80 percent of salary to allow a young researcher to work for two years part-time in an SME and part-time in a research lab while conducting a specific research project for the firm. Other criteria for access to a First-Enterprise project is that the researcher should spend enough time in the hosting research institution to allow a transfer of substantial knowledge to the firm, and that the research team in the institution must be competent in the relevant research field and motivated to engage in innovation projects for the firm (OECD 2004). Similar schemes, adapted to local conditions, could be useful to China as well. To the extent that the salary, welfare, and career development potential offered by SMEs to a technical expert are often significantly less attractive than what a large company or RDI can offer, personnel mobility schemes have great potential to improve the innovation performance of the private sector and justify government support.

Enhancing Innovation Services

Innovation, be it the creation or adoption of technology, is often a complex process that requires a wide range of complementary services. Enterprises, especially SMEs, often need to outsource such services to make innovation more effective and efficient. The list of such services can be long; table 3.12 provides some examples.

Issues
As with other services, there are two issues in the delivery of innovation support: financing and provision. Many innovation services are in the nature of a public good and require government financing, but most innovation

Table 3.12 Experience of Chongqing and Zhejiang SMEs with Innovation Services, 2006

| Service or service p | Region | Number of valid responses | Distribution of valid responses (percent) | | | | |
| | | | No demand | No access | Unsatisfactory | Results of service | |
						Largely satisfactory	Extremely satisfactory
Technical consulting/training	Chongqing	194	14.4	29.4	19.1	33.5	3.6
	Zhejiang	149	8.1	10.7	24.2	43.0	14.1
Information	Chongqing	186	17.2	28.5	19.4	32.8	2.2
	Zhejiang	149	6.7	15.4	24.8	40.3	12.8
Testing/processing center	Chongqing	184	15.2	25.5	14.7	39.1	5.4
	Zhejiang	146	3.4	17.1	24.7	36.3	18.5
Technology/IPR evaluation/transfer	Chongqing	182	25.3	31.9	17.0	23.6	2.2
	Zhejiang	148	10.8	23.6	18.9	35.1	11.5
Legal/patent	Chongqing	189	13.2	16.9	15.3	47.1	7.4
	Zhejiang	148	4.1	14.9	16.9	48.0	16.2
Headhunting	Chongqing	178	34.8	51.7	6.7	5.1	1.7
	Zhejiang	146	10.3	38.4	15.8	21.2	14.4
Finance/investment consulting	Chongqing	183	29.5	26.8	16.9	25.7	1.1
	Zhejiang	147	8.8	17.0	23.1	38.8	12.2

Credit guarantee	Chongqing	185	26.5	43.2	13.0	14.1	3.2
	Zhejiang	143	11.2	14.0	26.6	35.7	12.6
Industry association	Chongqing	185	15.1	27.6	13.0	37.8	6.5
	Zhejiang	148	5.4	11.5	20.9	44.6	17.6
Productivity center	Chongqing	186	15.1	39.2	13.4	23.1	9.1
	Zhejiang	140	10.0	30.7	15.0	32.1	12.1
Incubator	Chongqing	173	23.7	53.2	6.9	9.8	6.4
	Zhejiang	135	8.9	44.4	11.9	21.5	13.3

Source: World Bank CSMEI Survey.

Note: Respondents were asked to mark one of the following six items to describe their own experiences; 1 = My firm did not have this kind of demand; 2 = There is no this kind of service providers around; 3 = There is this kind of service providers around, but my firm never used the service because it is hard to get it; 4 = My firm used this kind of services but the results were unsatisfactory; 5 = My firm used this kind of services and the results were largely satisfactory; 6 = My firm used this kind of services and the results were extremely satisfactory. The responses are summarized with the following labels: No demand = 1; No access = 2 and 3; Results unsatisfactory = 4; Results largely satisfactory = 5; Results extremely satisfactory = 6.

services can be provided by nongovernment entities.[43] Moreover, the government can stimulate the development of the innovation services industry in the private sector, mainly through the creation of a favorable investment climate.

China already has a large number of innovation service providers—as of 2004, about 70,000 employing about 1.2 million people in mid- to large-size cities. Among these firms are 1,218 productivity promotion centers and 464 incubators (excluding 42 university high-tech parks).[44] However, empirical evidence suggests that the benefit that Chinese SMEs derive from this sector remains far from satisfactory. In the NBS survey, average ratings of the surveyed firms regarding the quality of various services providers are all under 3.5 on an ascending scale of 1–5. The World Bank CSMEI Survey confirms this general picture and provides some further details.

In the CSMEI Survey, respondents in Chongqing and Zhejiang were asked to describe their experience with 11 kinds of services or service providers (table 3.12). Overall, firms in Zhejiang were clearly more satisfied with innovation services around them, and more active in using them, than firms in Chongqing: 50.2 percent of firms in Zhejiang were "largely" or "extremely" satisfied versus 31.0 percent in Chongqing; and only 8.0 percent of Zhejiang respondents on average said they did not need the services in question, whereas the average was 20.9 percent in the case of Chongqing. The service providers receiving the best ratings from firms in both locations were legal and patent firms, industrial associations, and testing and processing centers. Headhunting and incubator services received the worst ratings, largely because of access difficulties.

The reasons for the general underdevelopment of innovation services in China are obviously complex. The various service sectors differ in their technical and economic characteristics, and so each requires individual analysis. However, the fact that Zhejiang firms expressed a significantly higher rate of satisfaction than did Chongqing firms calls for further investigation into the causes behind it. It is likely that the quality of innovation services is related to the general level of development of the private sector and market institutions, in which Zhejiang is widely considered as more advanced.

Recommendations

Although each service sector deserves its own assessment, the government and the private sector can consider some actions to promote the development of innovation services targeting SMEs, especially in inland regions such as Chongqing.

Strengthening government support for innovation services that provide public goods. Government departments and agencies could be given the authority to identify services that provide a public good, be permitted to include financing for them in their budget requests, and be held accountable for results. Most such services could then be purchased from nongovernment providers by responsible government departments and agencies through various means. For example, through subsidies, a local government could encourage local SMEs to purchase services from IPR agencies to increase the firms' awareness and knowledge of IPR. It could also pay for a private entity to develop and put in public domain a technical standard for a sector.

A rigorous monitoring and evaluation system should be established to measure the results. Where the needs for such public spending are relatively concentrated, a special fund could be created to manage the spending in a systematic manner. In some cases, the government may be justified in providing the services directly through publicly owned providers such as public service units. ITRI of Taiwan, China, and Fundación Chile are two successful examples in this regard (boxes 3.6 and 3.7).

It is worth noting, however, that the utility of service providers such as ITRI of Taiwan, China, (box 3.6) can be realized only when they have a clear objective—for example, exploiting opportunities and emerging markets that have a promising long-run potential. The providers must also be part of a strategy grounded in a solid assessment of existing (and relevant) resources and capabilities (skills and infrastructure); and part of a plan for remedying current deficiencies in skills, infrastructure, and research so as to meet the objectives.

Supporting MSTQ (measurements, standards, testing, and quality) services. Standards and quality are closely linked to innovation and productivity. Quality standards supported by a national MSTQ system can contribute to enterprise competitiveness, innovation, and trade. They do so by improving information flows and allowing customer differentiation, thereby promoting quality and enhancing competition. Standards also embody technology, thereby acting as a channel for technology diffusion and enhancing productivity (Dutz 2007).

China's MSTQ system, represented by the National Institute of Metrology, is dominated by PSUs. The government could consider further

Box 3.6

The Industrial Technology Research Institute (ITRI) in Taiwan, China

Publicly supported technological and scientific research institutions acting as innovation intermediaries have been indispensable in the high-tech industrial development of Taiwan, China. The Industrial Technology Research Institute (ITRI) and Electronics Research and Services Organization (ERSO), for example, based in Hsinchu, have played a major role in developing the technology capacities of local firms.

Established in 1973, ITRI employs more than 6,000 people, including nearly 5,000 R&D employees (820 with PhDs), and has an annual operating budget of $500 million. Its technology focus ranges from the high-tech integrated circuit (IC) industry to the textile industry, and its work on factory automation and advanced materials has also been applied in traditional industries.

ITRI coordinates industry consortia such as the Taiwan New Personal Computer (TNPC) Alliance, formed in 1993, which involved 31 partners including IBM, Apple, and Motorola. The aims of the alliance were to "bring together firms from all aspects of the IT industry with a clear focus on transferring, up-taking and diffusing the new PowerPC technology in a series of products spanning PCs, software, peripherals and applications such as multimedia" (Mathews and Poon 1995, pp. 43–58). The initiative behind TNPC lay with the Computer and Communications Laboratory, a part of ITRI.

Another example of ITRI's carefully developed role as an intermediary is its Open Laboratory Program, begun in 1996 and based in an extensive R&D complex in Hsinchu. The program mainly provides space and facilities for joint R&D between ITRI researchers and local businesses and also has space for business incubation, conferences, and training facilities. Firms in the incubator receive "packaged" business and management consulting, financial and legal assistance, and office and administrative support. Entry to the business incubator requires formal approval of a business plan. Such consultancy activity is subsidized by the government.

Sources: Dodgson, Mathews, and Kastelle 2006; Mathews and Poon 1995.

strengthening the contribution of this system to the capacity building of private SMEs in the following ways:

- Review the functioning of all MSTQ programs, including their governance and management structures and their effectiveness, with a view to improving their operational effectiveness and maximizing synergies between initiatives sponsored by various line ministries.

Box 3.7

Fundación Chile: A Path Breaker in Tapping Technologies and Promoting Innovations

Fundación Chile is a privately owned, nonprofit technology center created in 1976 with a mandate to develop innovative businesses and programs by transferring technologies. Its mission is to contribute to innovation in markets for goods and services and to transfer technologies aimed at providing Chile with added economic value. Fundación Chile focuses on improving the technical performance of economically important sectors by tapping advanced global technologies to create new companies and joint ventures.

Its technology transfer mechanisms include (1) R&D and adaptation of foreign technology for product and process innovation; (2) promotion of a technology "consortium for pre-competitive horizontal R&D"; (3) technological extension and technology diffusion to SMEs; and (4) institutional innovation (oriented to reducing transaction costs, the development of incomplete markets, and public-private partnerships for institutional development).

Fundación Chile identifies "missing links" necessary to provide specific industrial "clusters" with a comparative advantage. The clusters include the agribusiness, marine, tourism (agro/eco), forestry, and wood processing sectors. It also applies the missing-link concept in supporting education and human resources development.

Fundación Chile has been quite successful in incubating new ventures through entrepreneurship and technological innovation. By 1999 it had launched 36 such ventures; 17 have been sold. The six leading new ventures have generated more revenue than the total cost of the Fundación since its inception.

Source: World Bank 2004.

- Increase industry awareness of MSTQ services and their importance, including through better interaction with industry organizations and incentives (such as matching grants) for SMEs to use MSTQ services and obtain national and ISO certifications.
- Increase funding and staffing support for the nation's metrology infrastructure and encourage private participation in labs for testing and accreditation. Many government testing and accreditation labs could be considered for privatization or at least private management.
- Encourage participation by Chinese scientists and MSTQ personnel (public and private) in international technical committees, working groups, workshops, and seminars.

Reforming industry associations. Industry associations can deliver important benefits to its member firms, such as information services, policy and technology consulting, the organization of trade and investment opportunities, and training. They can also help build bridges between the government and the private sector. In China, however, industry associations traditionally have been an extension of the government and have had little autonomy. To fully realize the potential of industry associations and enable them to better serve the needs of the private sector, China needs to transform them into truly nongovernmental organizations by allowing firms to set them up freely and elect their managers. Certain policy incentives, such as tax benefits and competitive grants for nonprofit organizations, should apply to industry associations. The government can also support the training of association staff members to help them better perform their roles. A good example of an effective industry association is the Semiconductor Industry Association, in the United States (box 3.8).

Box 3.8

The Semiconductor Industry Association

The Semiconductor Industry Association (SIA) was founded in the United States in 1977 by five microelectronics firms. Its 95 member companies account for more than 85 percent of the U.S. production of semiconductors, an $80 billion industry.

The association strives to advance the global competitiveness of its members and their industry through a network of corporate leaders and working committees. SIA has 10 committees addressing industry and member concerns regarding trade, technology, occupational safety and health, environmental issues, industry statistics, government procurement, and other areas of industry and public policy.

SIA also provides services to directly aid its members. Among those services are (1) helping firms educate and recruit highly skilled employees, (2) providing advisory services in semiconductor technologies, (3) promoting fair and open trade, (4) helping provide safe working conditions in production facilities, (5) helping protect the environment, (6) tracking and distributing statistical information on market trends, (7) providing market research and policy consulting services, and (8) organizing trade and investment events for member companies.

Source: http://www.sia-online.org.

Conclusions

In China's existing national innovation system, state-owned LMEs and RDIs are the main performers of innovation activities. In the future, however, China's success in technological catch-up is likely to rely more on the capacity of its private sector. The past decade has witnessed a spectacular takeoff of China's private sector that has been fueled by dynamic business creation. Most Chinese private firms now are young SMEs run by inexperienced owners and managers operating with relatively low technology. For the private sector to play a leading role in innovation in China, building its capacity for technology absorption and creation is of strategic importance.

The World Bank CSMEI Survey conducted in late 2006 and early 2007 provides information on characteristics of the innovation activities of SMEs in two regions, the southwestern inland city of Chongqing, where the sample firms are more technology based, and the coastal province of Zhejiang, where the sample firms are mostly in traditional manufacturing industries. They share some characteristics while differing in some other ways (table 3.13).

Table 3.13   Comparison of SMEs in Chongqing and Zhejiang regarding Innovation Activities: A Summary

Item	Similarities	Differences
Why innovate?	To develop new markets and new products	Zhejiang firms emphasize the improvement of existing products, cost reduction, and complying with regulations and standards
How to innovate?	Build own R&D team	Zhejiang firms make more use of external resources by collaborating with suppliers, clients, and competitors and by taking over other firms
Cooperation with RDIs/HEIs	Technical consulting is the main mode	Joint research is more important to Chongqing firms; training is more important to Zhejiang firms
Difficulties experienced	Shortage of talent	Chongqing firms suffer more from economic limitations such as lack of capital and market demand; Zhejiang firms have more difficulty with technical capacity

Source: World Bank CSMEI Survey.

The perceived shortage of skilled workers is clearly a constraint on private SMEs, but it needs to be interpreted with caution—analysis of the data suggests that it is tied to inadequate internal HR management. Hence, both the private sector and the government need to invest more in improving HR management in private SMEs and are advised to consider the following actions:

- Modernize HR management, starting with enforcement of labor rights under the Labor Contract Law.
- Make use of confidentiality agreements and competition restrictions under the Labor Contract Law to protect technical secrets from being taken by employees when they resign.
- Facilitate better adaptation by SMEs to labor market conditions. In particular, local governments could create SME skill development centers to (1) provide SMEs with management and technical training especially related to innovation; (2) provide information on the demand and supply for various skills and the premium on various job categories through close relationships with schools, training institutions, and the labor market; and (3) collect and disseminate success stories, especially those from inside China, about the management of skilled employees and the promotion of an innovation culture.
- Strengthen policies supporting training and vocational education by reviewing the ceiling on the tax-deductible training expenditures of enterprises and redefining the role of the government in vocational education.

Improvements can also be made in facilitating the collaboration of SMEs with knowledge institutions and enhancing innovation services:

- The government could consider initiatives to facilitate SME participation in innovation networks. These may include, for example, innovation brokering programs following the model of the Norwegian TEFT, an innovation voucher program following the Dutch example, and personnel mobility schemes similar to those of the UK Business Fellowship.
- The government could promote the development of innovation services in specific sectors. In particular, it could further strengthen support of those innovation services that are of a public goods nature. Acceleration of the reform of industrial associations is also desirable.

CHAPTER 4

Strengthening the Ecosystem for the Venture Capital Industry

Innovation—whether achieved through adoption, adaptation, or the creation of new technology—needs to be financed. And financing the innovation activities of China's business enterprises is clearly an area where further improvement can be made. For example, as noted in the previous chapter, respondents to the World Bank CSMEI Survey ranked "lack of capital" as one of the top difficulties they face in pursuing innovation. Innovation financing is not just about allocating money for innovation. It is also about the effectiveness and efficiency of innovation activities, because performance in innovation can be enhanced by improvements in corporate governance demanded by a firm's investors.

Innovation can be financed internally or externally. External financing generally plays a positive role in the commercialization stage of an innovation. In early-stage technological development (ESTD), evidence in developed economies suggests that most innovative firms use internal and informal financing. In a basic framework for innovation financing, the types of financing used changes as innovation passes through various stages from invention to sale of the firm or product (table 4.1). In that framework, formal financial institutions, such as venture capital (VC) funds, private equity, investors, and banks, do not get involved until the early phases of production.

Table 4.1   Basic Framework for Types of Financing Used at Selected Stages of Innovation

Stage	Own funds, friends, and family	Angel investors, seed funding	Government programs	Corporate venture (retained earnings)	VC	Private equity	Banks, investment banks
Invention and R&D	√	√	√	√			
Business plan and market definition	√	√	√	√			
Pilot production		√	√	√	√		
Marketing, sales, and distribution				√	√		
Full commercialization					√	√	
Market expansion and increased penetration						√	√
Sale of company or product						√	√

Source: Based on Goldberg 2004.

Firms that are in the start-up or very early stage in bringing a new invention to the market typically experience the greatest difficulty in raising funds. The difficulties of early funding are even more pronounced in emerging markets, where the formal financial institutions that provide various forms of risk capital are not well established. In the United States, one of the world's most advanced and innovative economies, ESTD is largely financed by an entrepreneur or the firm's own funds and by government programs (figure 4.1). Internal funding overcomes the problem of information asymmetries with regard to the quality of the innovation, potential market applications, and commercialization, but internal funding is available only to entrepreneurs and firms with sufficient cash. Government funding comes in a variety of forms, but it is mainly grant based; and given the public policy nature of such funding, the asymmetric risks and potential for return are not necessarily the overriding concerns. The other main providers of early funding in the United States are "angel" investors. Angel investors are generally successful entrepreneurs that invest in ESTD projects in fields they have already succeeded in, and these investors generally get deeply involved in the business development and management of the investee firms. In the

Figure 4.1   Sources of Funding for Early-Stage Technology Development in the United States, 2002

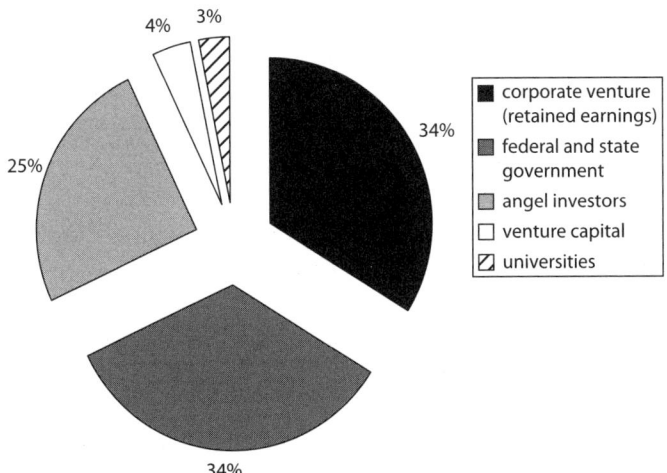

4%   3%

34%

25%

corporate venture
(retained earnings)

federal and state
government

angel investors

venture capital

universities

34%

Source: Auerswald and Branscomb 2003.

United States, financial institutions in the ESTD phase of innovation provide very little external risk capital—in 2002, VC funds provided only 4 percent of all ESTD funding (Auerswald and Branscomb 2003).

As entrepreneurs and firms move from ESTD toward commercialization, a number of factors still make financing a challenge: low expected returns due to an inability to capture the profits from an invention or innovation, the uncertainty and risk associated with the project, and overoptimism and untested capacities on the part of managers (most of whom may not have any tangible assets or track records) (Hall 2005). Some of these risks can be mitigated through intellectual property protection, subsidies, or tax incentives, but there is still a "wedge, sometimes large, between the rate of return required by an entrepreneur investing his own funds and that required by external investors" (Hall 2005, p. 3).

The combination of risks, both perceived and actual, surrounding innovation makes its financing stand out as a special issue because it requires not only capital, but also external risk capital—that is, capital from external providers that are willing and able to take high risks involved in the creation, adaptation, and adoption of technology. This is where the VC industry emerged to fill the funding gap, or wedge, for start-up and early-stage risk capital for innovative firms.

Of course, VC does not completely fill the financing void for innovative firms, as was shown earlier. However, research has shown that VC plays a strong role in encouraging innovation and, more important, facilitating the commercialization of innovation. VC firms play their role in several ways: expert selection of investments, expert advice to investee firms, assistance in business management and corporate governance, connecting of firms to potential buyers of their products, efficient longer-term financing of the firms, and performance monitoring and advice from the start to the realization (sale, or "exit") of the VC investment (Lerner and Watson 2007). International experience also shows that VC is a key area in which governments commonly intervene in attempting to catalyze new and early-growth innovative businesses.

This chapter studies the emerging domestic VC industry in China, drawing in part on Mackenzie (2007). It explores ways in which the development of the industry can be catalyzed to support the enterprise-led innovation strategy. The discussion here emphasizes a view of the VC industry as an ecosystem.

The Ecosystem for the VC Industry

VC has a long history dating back to the period of the 15th to the 18th centuries, when the rulers of Europe supported the exploration and subsequent colonization of many parts of the world.[45] Modern-day venture funding emerged in the United States in 1958 with the creation of government-backed Small Business Investment Companies and subsequently with the development of private VC funds in the 1960s and 1970s (Lake and Lake 2000, p. 6). The U.S. VC industry has gone through a number of activity cycles, and the most recent boom began in 2001. The evolution of the VC industry has shown that its success depends on a surrounding "ecosystem." The VC ecosystem has many elements, but four are fundamental: structure, funding, management, and exit.

Structure of a VC Fund

A VC fund, structured as an independently managed pool of funds from institutions and individuals, is generally managed by a VC firm. The VC firm invests the funds in firms to support them in four basic stages of development: (1) seed or start-up, (2) early growth, (3) business expansion, and (4) later-stage activities.[46] The investment essentially represents a business partnership in which the VC firm shares the risks and rewards of the business and provides advice and expert assistance. The United

States is the world's largest private equity and VC market, and therefore the legal form of the industry has been shaped by the U.S. experience. The primary legal form of VC firms, with local variations, is usually a fixed-life investment vehicle that consists of a general partner, or GP (the management firm, which has unlimited liability), and limited partners, or LPs (the investors, who have limited liability). In managing the partnership, the GP receives a management fee (usually 2 percent) and up to 20 percent of the profits (sometimes called "carried interest").[47] The LPs receive income, capital gains, and tax benefits from their investments. The fee structure helps to align the interests of the GP with LPs to maximize returns while covering the minimum operating costs to run the fund.

The LPs generally invest in the fund without knowledge of which companies the fund will invest in (a "blind pool"), and they do not participate in the investment decision-making process or operations of the fund. The GP usually has actual or effective control over the portfolio companies and specializes in "finding, analyzing, investing in, managing, and exiting in what are generally private companies" (Lake and Lake 2000, p. 10). The GP usually has expertise in a particular sector (for example, information technology, biotechnology, or health care), in the markets of a particular country, and in one of the four investment stages (seed, start-up, and so on). The partnership is a closed-end fund that normally lasts 10 years, during which time the investment is essentially locked-up and illiquid (although technically the investor can sell the partnership interest).

The reason for the evolution of a special corporate form for VC funding, the limited partnership, is the need to balance the desire of the investor to incentivize the VC firm to make attractive returns with the desire to restrict the VC firm from engaging in unduly risky behavior or from engaging in activities that would conflict with the investors' objectives. The VC fund by definition invests in risky and illiquid companies that will in most cases take many years to bring to profitability and liquidity; the VC firm therefore needs the time and freedom to pursue that approach without undue pressure from investors for short-term profits while still being under pressure to generate profits for the investors in the medium to long term. Therefore, the typical corporate form of a limited liability or joint stock company, which normally is of unlimited duration and in which the investors ultimately control the company via the board of directors or shareholders' body, has been seen as not ideal for VC investing. (Gompers and Lerner 1999; Lerner 2000). Finally, limited partnerships have the added benefits of being tax efficient because usually the limited partnerships are nontaxable entities—profits generated by the fund are not taxed at the limited partnership level. They flow back to the individual investors, who are taxed on the basis of their own individual situation, and, thus, profits avoid double taxation. A number of legal jurisdictions around the world with investor-friendly corporate and tax laws are the most popular for domiciling VC limited partnerships, including the United States, the United Kingdom, and the Cayman Islands.

Sources of VC Funding

VC firms are intermediaries whose funding is derived from institutional and individual investors (the latter are sometimes referred to as "family offices"). Institutional investors include pension funds, banks, investment funds, and other specialized financial institutions. Also, some corporations or financial institutions establish their own VC vehicles, but those entities represent a much smaller portion of the VC industry. In most developed countries, such as the United States and in western Europe, institutional investors have many decades of experience in VC investing. There is a deep pool of funding for VC investments, but VC will generally represent only a small portion of an investor's portfolio, given the high level of risks in this market segment.

The higher returns and long-term nature of VC investments are attracting an increasing amount of funds into VC. For example, in the United States, most large institutional investors now tend to allocate 5–15 percent of their total assets in alternative investments, which mainly consists of hedge funds, private equity, and VC investments. In Europe, pension funds provide 25 percent of all funding for VC (and private equity), followed by banks (16 percent), funds (13 percent), and insurance companies (10 percent); and the majority of the funding for the industry comes from within Europe (63.4 percent).[48]

The Role of VC in Investee Companies

VC firms need to operate under legal guidelines that allow the investee companies and the investments to be structured to align the interests of VC funds and the companies they invest in.

In the basic VC investment situation, a small team of entrepreneurs has a new technology or idea around which a business can be built but does not have the capital to fund it. A VC fund that is attracted to the project provides some initial equity investment as well as expertise in developing and commercializing the idea. The successful VC firm typically has management and staff members with entrepreneurial experience, sector experience, or other specialties that can be applied directly in the investee company. As previously mentioned, the VC firm is essentially a business partner of the investee company and typically is highly involved in all decisions facing the investee firm, including the hiring and firing of management, and actively participates in the governance of the company. The VC firm also gets deeply into the strategic planning of the investee company and into some operational areas such as product development, distribution, and marketing. Another valued skill brought by the VC firm is a network of contacts to help the investee company make links with suppliers and buyers and with other investors to finance the next stage of growth (Dotzler 2001). Therefore, the VC industry is necessarily largely a localized business because the VC firm must have in-depth knowledge of the management and market of the investee company in order to add sufficient value to bring the company to the next stage of development.

VC investments are generally made with special classes of shares and with numerous controls and conditions that allow the investor to have a degree of preferential treatment and control that would not exist if both the investee team and the investor had the same class of shares. The reason for the preferential status is the high degree of risk surrounding investments in innovative sectors, including lack of proof of concept of the business idea, information asymmetry between the investor and investee, and a lack of assets at the investee company. Generally, during the life of a typical VC-financed company, the capital structure will be continuously changed, a process requiring a very flexible regulatory environment.

Exit: Realizing VC Investments

On the investor side, a VC fund offers new channels for diversifying risks and opportunities for higher returns as well as a longer-term investment vehicle. The exit strategies for a VC investment generally fall into one of four categories if the investment is successful: (1) initial public offering (IPO), in which the shares of the invested firm are sold to the public on a stock exchange, (2) merger and acquisition (M&A), in which the shares are sold to a third party that is typically a strategic buyer, (3) buyback, in which the firm's management buys the company's shares from the VC investors, and (4) sale of the firm. Of course, if the investment is a failure, the scenario is different, and other options (such as a write-off) constitute the investors' choices for exit.

Exit via an IPO is generally the most lucrative option, but the firm for sale must meet many prerequisites, such as long-term stability, persistent profitability, sound cash flow, strong customer base, sizable market share, good corporate governance compliance, and effective management. Recent surveys of U.S. venture funds indicate that the preferred exit route changes on average 1–3 times during the life of the investment (Center for Private Equity and Entrepreneurship 2005). However, in the United States, whose VC industry is the largest and most mature in the world, sale to a strategic buyer is overwhelmingly the preferred exit strategy, particularly early in the life of the investment. Although the average size of VC-backed IPOs was $120 million in 2007, which was larger than M&A exits ($78 million), exits via M&A were 3.5 times more numerous than via IPO. In addition, M&A exits represented 70 percent of the total value of all VC-backed divestments in the United States (National VC Association 2007). In Europe, the dynamics are very similar, but in emerging markets across Asia, the situation appears to be exactly the opposite—exits are dominated by IPOs. For example, in 2007, 78 percent of the total number of divestments in Asia were in the form of IPOs (Centre for Asia Private Equity Research 2007). However, the dominance of IPOs in Asia probably reflects the underdeveloped environment for M&As and the fast-growing stock markets in the region more than anything else.

The failure rate of individual VC investments is quite high, with some estimates ranging up to 40 percent (Ernst and Young Venture Capital Advisory Group 2006). One VC fund in the United States estimates that the experience with the total portfolio of investee companies is largely broken down into thirds, with one-third of the companies failing, one-third underperforming expectations, and one-third meeting expectations and generating a return of five times the original investment (Union Square Ventures 2007). Therefore, a VC fund must generate a large flow of deals and high returns from the investee companies that succeed. Despite the high failure rate, the overall industry has yielded high returns relative to benchmark equity indexes. By the end of 2007, VC funds in the United States provided a 10-year annualized return of 18.3 percent, with seed and start-up funds providing a return of 35.5 percent, versus a return of 5.3 percent for the Nasdaq and 4.2 percent for the S&P 500 stock indexes (National VC Association 2008). Due to the need for such high rates of return, VC funding tends to go toward clusters of innovation located in particular regions and into sectors with particularly high growth potential. In the United States, the life sciences industry (biotechnology and medical devices) captured 31 percent of all VC investments in 2007, followed by the software industry (18 percent), and the energy sector (9 percent). VC also tends to be clustered in the locations where those economic activities are based, so 34 percent of all U.S. VC investments in 2007 went to Silicon Valley (California), 22 percent to New England, and 7 percent to the San Diego area (National VC Association 2007).

The Domestic VC Industry in China

From its inception, China's domestic VC industry has had heavy government backing. In 1984 the importance of VC was officially recognized, and in 1985 the first VC firm, China VC Company for New Technologies, was established with government funds (and was subsequently closed down in 1997) (Bottelier 2004). Since the establishment of the first VC companies in Shenzhen in the late 1980s, governments at all levels in China have invested directly in VC funds as majority shareholders or directly in start-up high-tech firms. Domestic VC funds in 2006 raised 37.2 percent of their funding from the government, SOEs, and PSUs (figure 4.2).

The government-sponsored VC funds have traditionally targeted either specific industry sectors or certain types of firms, such as SMEs. One example of that type of fund is the National Electronic and Information Technology Development Fund (IT Fund), sponsored by the former Ministry of Information Industry (MII). The fund makes equity investments in high-growth, technology-based SMEs in the information

Figure 4.2   Sources of Funding for China's Domestic VC Firms

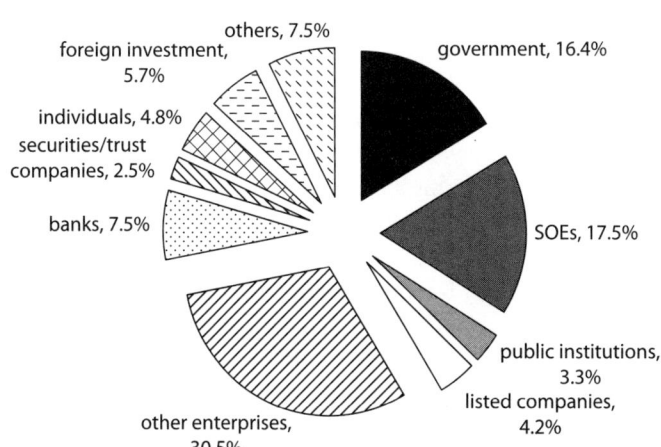

Source: Wang, Wang, and Liang 2007.

technology sector. By the end of 2006, the IT Fund had invested Y 266 million in 24 projects, with a total rate of return of 86 percent over eight years and total value of equity of Y 497 million.[49] One highly visible investment by the IT Fund was in China Vimicro, a small firm developing integrated circuits in Zhongguancun, Beijing, by a team of overseas returnees. With an investment of Y 10 million, the IT Fund owned 12 percent of the company, which was then charged by MII, through the IT Fund, to develop China's digital microchip industry and design commercial products. By the end of 2006, Vimicro became an internationally competitive producer of multimedia processors, with more than 100 million chips sold in 16 countries and regions. Vimicro was listed on the Nasdaq on November 15, 2005.

In some cases, local governments either set up VC funds or invest in private sector VC funds (rather than in high-tech firms directly) to attract more private investment into high-tech firms. Examples include the Zhongguancun High Tech VC Co-financing Fund, Suzhou Industrial Park VC Co-financing Fund (Y 1 billion for the first round), and Shanghai Pudong New Economic Zone VC Co-financing Fund (Y 1 billion from the local government and Y 1 billion from other sources). Meanwhile, the Ministry of Finance (MOF) approved a Y 100 million National VC Promotion Fund in the government's budget in early 2007 to nurture the development of a robust domestic VC industry. That government VC fund claims to be modeled after Israel's Yozma Fund.[50] The National VC Promotion Fund is designed to be a "fund of funds," investing in other VC funds instead of directly making investments in companies so as to attract more private VC investment in the early stage for high-tech firms. As of this writing, the fund has made no investments. Very little information is publicly available on the direct role of the government at any level (national, provincial, or municipal) in the VC industry; therefore, verifying the data and assessing the effectiveness of government intervention is not possible at this stage.

The direct role of the government in the VC industry was recently detailed in a document jointly issued by the MOF and MOST on July 6, 2007: "Interim Measures for the Administration of Guidance Funds for Promoting VC Investments in Small and Medium-Sized Technology Enterprises." As they are currently understood, the so-called interim measures apply to all levels of government. The document covers the purpose, funding, investment requirements, and organizational structures of "Guidance Funds," through which government will direct and promote VC investment in high-tech SMEs at the start-up stage of their business.

The Guidance Funds may invest in VC funds or directly in target companies.[51] Capital for the Guidance Funds would be allocated from the central government and would be invested with VC firms with a minimum capitalization of at least Y 100 million. The Guidance Funds could be invested (1) through equity investments in VC funds, (2) through coinvestment with a VC fund in a particular targeted SME, (3) directly in a targeted SME as seed capital, or (4) in the form of an initial loss-funding for investments in the targeted high-tech SMEs. The interim measures thus allow for a wide range of activities by the government at all levels in the VC space in China.

The Entrance of Foreign VC Funds

The early history and statistics suggest that government was the dominant player at the outset of the Chinese VC industry. Although the government still has a strong presence in the domestic VC industry, foreign VC firms have come to control the majority of VC activities in China today. In 1992, the first foreign VC fund was established in China by the International Data Group (IDG). By 2007, foreign VC funds had raised 82 percent of all new VC investment by value, with only 13 percent coming from domestic VC funds (table 4.2). Because the number of new VC funds that year was split fairly evenly between domestic and foreign, the average amount of capital in the new foreign funds was more than five times that in the new domestic funds.

Foreign VC funds were also more active in terms of investing in China than were their domestic counterparts (table 4.3). Investment by foreign VC funds in 2007 represented 89 percent of the total value of all new VC investments in China in 2007. The average deal size for the domestic firms was about one-third that for the foreign firms. The average size of joint venture deals was about equal to that for foreign VC funds.

Table 4.2   Number of New VC Firms Started in China, and Amount of Capitalization, by Domestic and Foreign Origin, 2007

		Capitalization		
Origin of VC funds	Number of new funds	Value ($ millions)	Average per fund ($ millions)	Share of value (percent)
Domestic	25	1,106.21	44.25	13
Foreign	29	6,886.72	237.47	82
Joint venture	4	437.71	109.43	5
Total	**58**	**8,430.64**	**145.36**	**100**

Source: Zero2IPO 2007.

Table 4.3   Investments in China Made by VC Funds, by Domestic and Foreign Origin, 2007

Origin of VC funds	Number of investments	Value ($ millions)	Average value of investment ($ millions)	Share of total value (percent)
Domestic	87	290.53	3.34	8
Foreign	317	3178.68	10.03	89
Joint venture	11	119.61	10.87	3
Total	**415**	**3,588.82**	**8.65**	**100**

Source: Zero2IPO 2007.

These differences are part of one of the most striking characteristics of VC activities in China—the dual VC structure (figure 4.3). The essential differences between the two models, i.e., the foreign model and the domestic model, are clear. In the case of the foreign model, all of the activity, except for that of the ultimate operating company, takes place offshore, while in the domestic model all the activity takes place within China. In almost all cases, the vast majority of the capital invested by the VC fund in the offshore holding company is actually used to invest in the operating company—that is, the funds do flow into China.

Foreign VC funds have backed a number of successful innovative companies in China that were started by a new generation of Chinese entrepreneurs. Although a few of those companies have significant exports, most are focused on the domestic market, and in all cases, the vast majority of the value-creating activity occurs within China. All of the VC firms invested their funds via an offshore-registered holding company. They were all incorporated outside of China, and all of the institutional investors in these VC funds were foreign. But significantly, most of the staff members at the foreign VC firms involved in the deals were Chinese. All of these successful innovation companies successfully executed an IPO on a stock market. But in each case, it was not a domestic Chinese entity that was listed on a Chinese stock exchange; rather, it was the offshore holding company that was listed on a foreign stock exchange, specifically a U.S. exchange and mostly on the Nasdaq. Thus, although the businesses of these companies are in China, virtually all parts of the VC system that supported them were offshore (Mackenzie 2007).

Current VC Market Dynamics

The VC industry in China has grown by almost 150 percent in five years as measured by assets under management, moving from $11.3 billion in

Figure 4.3   The Dual VC Structure in China

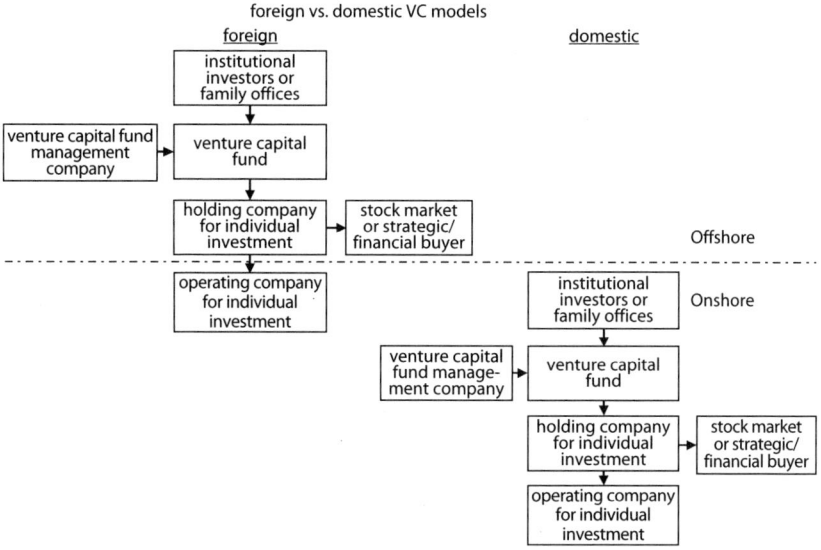

Source: Mackenzie 2007.

2003 to more than $28 billion by 2007 (table 4.4). The average annual growth in the value of investments was 47 percent during this period, and the average annual growth in the number of investments was 20 percent. The number of VC firms increased about 80 percent over those years, from 166 to 298, with 2005 witnessing the most rapid rise. The number of investments made per year rose 150 percent, from 164 in 2003 to 415 by 2007, and the average size of investments increased from $4.6 million per deal to $8.7 million.

In terms of the investment targets, most have been in the expansion stage of financing. Of the 415 new investments in 2007, 142 were for business expansion and constituted 54 percent of the total value of all venture investments (table 4.5). Developmental financing was in second place. However, early-stage financing still accounted for 22 percent of all deals that took place in 2007.

The growing capital markets in China have opened a domestic exit channel for VC-backed investments. IPOs represented 88 percent of all divestments in 2007, with M&A taking up a distant second place with 12 percent, a dramatic change from 2006, when IPOs were only

Table 4.4   Characteristics of VC in China, 2003–07

Characteristic	2003	2004	2005	2006	2007
Level					
Investments					
Number	164	194	298	362	415
Value ($ millions)	746.2	873.6	1,430.4	2,181.4	3,588.8
Average size ($ millions)	4.6	4.5	4.8	6.0	8.7
VC firms					
Number	166	183	256	278	298
Capital managed ($ millions)	11,310	11,870	17,210	20,042	28,043
Change (percent)					
Investments					
Number		−8	18	54	21	15
Value	40	17	64	53	65
Average size	52	−1	7	26	44
VC firms					
Number	9	10	40	9	7
Capital managed	8	5	45	16	40

Source: Zero2IPO 2007.

Table 4.5   VC Investments in China, by Stage of Business Development, 2007

Stage	Number of investments	Share of total number (percent)	Value ($ millions)	Share of total value (percent)	Average value of investment ($ millions)
Early	91	22	278.95	8	3.07
Development	163	39	1,160.83	32	7.12
Expansion	142	34	1,935.55	54	13.63
Profitable	19	5	213.49	6	11.24
Total	**415**	**100**	**3,588.82**	**100**	**8.65**

Source: Zero2IPO 2007.

43 percent of all divestments (table 4.6). Much of the change could be attributed to the confidence and blistering growth in the domestic capital markets in China in 2007, which in large measure was due to a series of important reforms in the capital markets (such as the nontradable-share-reform process).[52] While divestments grew only 10 percent from 2006 to 2007, the growth in divestments via domestic IPOs was 250 percent, and the 68 percent shrinkage in the use of M&As was an extraordinary change in divestment tactics by the VC industry in just one year.

An important reason for the improved environment in the domestic capital market was the creation of the Shenzhen SME Board in 2004.

Table 4.6   Number of VC Divestments in China, by Method, 2006–07

Method	2006	2007	Change (percent)
Initial public offering	43	96	123
Domestic	10	35	250
Foreign	33	61	85
Mergers and acquisitions	41	13	−68
Other	15	0	−100
Total	**99**	**109**	**10**

Source: Zero2IPO 2007.

Although it got off to a relatively shaky start, its performance has surged as the overall capital market has improved. As of year-end 2004, the Shenzhen SME Board had 38 listed companies with a total market capitalization of Y 41 billion. By year-end 2007, it had 202 listed firms with a total market capitalization of Y 1.1 trillion, a fourfold increase in listed firms and an increase in market capitalization of 25 times.[53] The number of listed companies has increased, and they include more than a dozen firms invested in by local VC companies. The debut of International Data Group–backed YGSOFT Corporation was especially in the spotlight. It was the first time that a venture backed by a foreign VC fund had debuted in China's domestic capital market.[54] Domestic VC funds had invested in 10 private Chinese companies that became listed on the Shenzhen SME Board in 2006, but those IPOs represented only 23 percent of all VC-backed IPOs in 2006—the rest of the transactions took place in foreign markets. However, progress was made in 2007, when 36 percent of all VC-backed IPOs (or 35 out of 96) were executed domestically (Zero2IPO 2007).

A final characteristic of the VC industry is financial performance. Actual performance of China's VC industry is difficult to estimate because performance measurement based on the available sources is not entirely clear or standardized. It would appear that those VC-backed firms that listed on both international and domestic exchanges yielded the highest returns to investors, with some IPOs returning many multiples (on the order of 4 to 10 times) of the original investment (as compared with about 5 times in the United States). But the failure rate for individual investments was relatively high, with some estimates suggesting upward of 60 percent of all investments (as compared with 20 percent to 40 percent in the United States) (Wang, Wang, and Liang 2007). Thus, it would appear that the performance of VC activity in China is governed by the extremes of both high risks and high returns.

Strengthening the Ecosystem for the VC Industry

Despite the rapid evolution and growth of VC in China, if faces many challenges that constrain China's ability to fully realize its potential in developing its innovative industries. In particular, with the exception of a few spectacular success cases, the domestic VC firms are still at an early stage of development, beset by a range of difficulties. The experience of foreign VC firms suggests that, contrary to conventional belief, the flow of innovations, ideas, and products within China that are worthy of VC investment does not seem to be the binding constraint. The further development of the domestic VC industry could benefit greatly from actions aimed at a strengthened ecosystem, particularly in four dimensions: structure, funding, management, and exit.

Testing of the New VC Structures

There are no major constraints to structuring a VC fund in China at this moment. The reforms in the corporate legal structures in China in the past three years represented a solid step forward (box 4.1). The revised Company Law and Partnership Law, as well as the subsequent regulatory issuances, effectively removed legal barriers to the proper structuring of a domestic VC fund in the form of a limited partnership. A few new funds reportedly have been established under the new laws, which will allow for the structure to be tested.[55] However, the revisions of the laws are so new that too few funds have so far been established under them. Therefore, the immediate challenge will be for some new funds to be established to allow for a domestic model to emerge.

Recommendations

With the close involvement of institutional investors, the government could conduct an assessment of the operations of the first batch of domestic VC funds created following the newly amended Partnership Law and identify loopholes and weaknesses that might require further legislative or policy actions. Once the weaknesses in the application of the legal framework for fund structuring have been identified, they should be addressed through new regulations.

Expanding of the Sources of Venture Funding

The potential sources of funding for the VC industry in China, such as investment funds, pension funds, banks, and insurance companies, are growing rapidly. The story of the size of China's banking system is well

Box 4.1

Recent Progress in the Reform of Company and Partnership Laws in China

In 2007, the National People's Congress ratified a substantial revision to China's Partnership Law, to take effect on June 1, 2007. The revisions make substantial improvements to the 1997 Partnership Law and have significant implications for the VC industry. The new Partnership Law allows for the following innovations: limited partnerships with two kinds of partners, general and limited; "pass-through" tax treatment—that is, the limited partnership is not a taxable entity, and profits and losses are passed through to the individual investors; legal persons are allowed as investors; and foreign natural persons and foreign legal persons are permitted to invest in domestic limited partnerships.

Also in 2007, the Ministry of Finance and State Administration of Taxation released a Circular Concerning the Tax Policies for the Promotion of Venture Capital Enterprises that granted properly approved venture capital enterprises making qualified venture capital investments the ability to deduct 70 percent of the amount of each individual investment from the income tax payable of the venture capital enterprise. This helps to address the residual issue of taxability of limited liability companies.

Source: Mackenzie 2007.

known, and financial institutions now have more than Y 40.1 trillion in total deposits as of year-end 2007 (PBOC 2007). The investment fund sector, which is relatively new by comparison, is much more modest in size, with about Y 3.1 trillion in assets under management in China at the end of 2007 (CSRC 2008). Although China's pension fund system is still very much a fledgling (and China has essentially no private foundations or endowments), its existing pension funds and insurance company assets are already quite substantial, and by most accounts, these pools of contractual savings will grow over the next few decades to be among the largest in the world. The National Social Security Fund (NSSF) has about Y 300 billion in assets.[56] Provincial government pension funds have an estimated Y 800 billion in assets.[57] Finally, total insurance company assets as of the middle of 2007 were in excess of Y 2.5 trillion.[58]

However, these funds are not making it into the VC arena. As was shown in figure 4.2, domestic VC funds receive no investment from insurance companies, pension funds, or other institutional investors. Only 7.5 percent of total funding is from banks, and 2.4 percent is from trust companies. The issue with capital from pension funds and insurance companies in China concerns investment policy and is quite simple: currently, China's pension funds and its insurance companies are prohibited from investing in private equity or VC funds, that is, they have zero percent of their funds allocated to this asset class.[59] It is obviously hard to develop a domestic VC industry if the major holders of domestic long-term capital are not allowed to invest in VC. Not only are the regulations problematic, but even in the absence of these regulations, there is a general perception in the market that VC funds are poorly managed, which further compounds the aversion to this asset class. The VC exposure of the more conservative investors, such as banks, is likely to have been the most affected by this risk perception. As for family offices, foundations, and endowments, these types of professional investors are essentially absent from China.

Recommendations

The main objectives of reform in this area would be twofold: first, to allow China's existing but nascent institutional investors to eventually begin investing in domestic venture capital institutions, and second, to promote the creation of other forms of investors oriented to the long term, such as family offices, foundations, and endowments. In the long run, China's institutional investors, such as its pension funds, insurance companies, investment funds, and banks, will likely emerge as the main source of funding for a truly viable domestic VC industry. The availability of such funding has been the case in economies with successful VC industries, where institutional investors lead the funding in VC as part of their effort to find higher long-term returns and to diversify their portfolio risk. In making their VC investments, these institutional investors are charged with delivering an appropriate risk-adjusted return while not putting the funds' principal at undue risk. The stringent regulatory limits placed on the various institutional investors in China on their exposure to VC as an asset class is, therefore, not conducive to the industry's development. The government should consider policy measures to allow for institutional investors to begin investing in domestic VC institutions.

Recognizing that the risks of VC investing are higher, the financial supervisory authorities may want to take a slow approach on this issue. The first step could be to develop a short- to medium-term action plan that would provide a roadmap allowing these institutional investors to invest in private equity and VC funds. One immediate first step could be to set new prudential guidelines on the investment by institutional investors in the wider alternative-asset class (hedge funds, private equity, and VC). The guidelines could set appropriate exposure limits and risk weightings for investments in alternatives that would be determined by the supervisory authorities in consultation with market participants.[60] As part of this gradual approach, the supervisory authorities could first allow only a select group of financial institutions to pilot such investments, provided that they qualify under some predefined criteria for measuring financial health and the quality of risk management. The supervisory authorities could allow these institutions to invest in established foreign private equity and VC funds only. Those funds could be foreign or domestic in terms of their investments to provide maximum flexibility and comfort to the regulatory authorities. This approach would give the institutions experience in investing in this asset class and allow them to learn how to pick fund managers.

Another step could be to allow institutional investors to support foreign private equity and VC fund managers that were willing to establish an RMB-denominated and China-domiciled fund. That would allow Chinese institutional investors to gain experience in investing in domestically oriented VC funds without taking excessive risk on new fund managers. An assessment could then be made of the risks, performance, and other factors, and the limits could then be adjusted upward by the authorities if warranted. In parallel, the financial supervisory bodies may also consider studying the regulation of the alternative-investment class in China with a view toward promoting its growth while maintaining a degree of transparency and stability.

Finally, but as a second-order priority, the government could promote the creation of other forms of long-term-oriented investors such as family offices, foundations, and endowments. As China's economy continues to grow rapidly, more and more entrepreneurs will become wealthy and will become potential VC investors. The government should, therefore, study the current and projected population of these investors and the detailed constraints they face in forming foundations, endowments, and other such vehicles.

Building Stronger Venture Partners for Investee Companies
The lack of professional investment fund managers in China is a particularly acute problem in the domestic VC industry. Given the relatively recent growth in the industry and the dominance of foreign VC funds, it is not surprising that few VC fund managers have a long track record of

investment. The domestic VC industry does not appear to have provided adequate incentives in terms of pay and promotion, particularly when compared with their foreign counterparts, which may have hindered them from attracting the best talent in the market. However, skills are even more necessary in the VC industry than in most other types of funds because of the nature of the investments—they require a hands-on approach and a range of specialized talents to extract value from the investee companies. The problem is compounded by a general dearth of high-quality business managers, and good and experienced new venture managers are a scarce resource in China (Deloitte Research 2006). Thus, such companies need even more help from VC investors at a time when the VC firms do not have sufficient numbers of skilled management and staff.

Inexperience has also led to problems in the corporate governance of domestic VC funds, especially for the large number of government-sponsored or -dominated VC. Despite significant variation across firms and regions, many of them suffer the same limitations as other state-owned financial institutions—the managers are largely government bureaucrats with mixed incentives and limited knowledge of VC investing, and they lack adequate risk-management controls. These governance problems extend to the wider corporate sector, where the culture and regulatory structures of developed economies, such as those of the United States and Europe, are not yet strongly rooted in China. Despite the new provisions in the Company Law that impose a duty of care and loyalty on directors, supervisors, and the senior management of companies, many ambiguities remain as to how directors would comply with the new requirements. In addition, many entrepreneurs do not fully understand the differences in institutional rights between VC, private equity, bank loans, or other forms of financing (Deloitte Research 2006).

In the most advanced legal jurisdictions, VC investors have the basic freedom to obtain the economic and governance rights appropriate to the risk profile of their investment. In addition, they have the basic freedom to restructure ownership of the company simply via registration and without government approval. The revised Company Law in China has introduced the concept of share classes; and even before the revisions, a company and investor faced no barriers in constructing a legal agreement to ascribe various rights. However, the problem is largely that in the case of a dispute, investors do not have adequate assurance that China's legal system would recognize such agreements even if they conform to the law; and enforcing contracts in China is time-consuming and uncertain given the state of the judicial system. Therefore, the exertion of control and consequent engagement in the key management decisions in an investee company by VC firms is quite challenging in China. The combination of inexperience and the weak corporate governance environment has resulted in some domestic VC funds acting as ordinary equity investors with a short horizon for their investment returns, a low tolerance of risks, and an inability to pick and choose truly profitable innovative firms and products.

Recommendations

The key challenges in this element of the VC ecosystem are long term. The lack of professional VC fund managers will evolve over time. Given that the domestic VC industry is nascent, the likely evolution of the industry will involve the transfer of knowledge from foreign VC firms to domestic firms. In that process, domestic firms could benefit from a clear talent strategy defined in connection with the overall business strategy of the firm, including, for example, market positioning. The best strategy is not necessarily one that intends to recruit and retain the best talents in the market. Rather, it is one that best fits the firm's overall strategy.

Improving corporate governance is another long-term challenge that is critical to strengthening the ecosystem of VC. In this regard, the primary policy actions that can be taken are related to enforcement of the amended 2005 Company Law. First, it is advisable for the government to organize the formulation of what might be called a "Code of Conduct for Corporate Governance" covering both state-owned and private companies. It can be an updated and expanded version of the Listed Company Corporate Governance Rules issued jointly by the China Securities Regulatory Commission (CSRC) and the State Economic and Trade Commission in January 2002.[61] In addition to general guidelines, the code of conduct should seek to operationalize the rules of the 2005 Company Law (box 4.2). For example, it could specify in more detail the concepts of duty of loyalty and duty of care and clarify the circumstances in which a director or officer can be viewed, according to the Company Law, as taking "advantage" of their "positions in the company" to exploit "opportunities that belong to the company," which also require specification. Second, to promote the proper use of preference shares, as in other developed company law environments such as those in the United States and Cayman Islands, the State Council might consider, as authorized by the Company Law, the promulgation of a regulation to govern the issuance of preference shares. The regulation could spell out,

Box 4.2

The 2005 Amendments to China's Company Law

Two amendments to China's Company Law, enacted in 2005, are particularly instrumental for further improving corporate governance in the context of VC investment.

The first was an explicit introduction of the concepts of "duty of loyalty" and "duty of care." Article 148 states that directors, supervisors, and executives owe a duty of loyalty and duty of care to the company. Although the scope of the duties is not further spelled out, article 149 provides a list of behaviors in which directors, supervisors, and executives "shall not" engage. Many items on the list are additions to the 1994 Company Law. For example, one item explicitly deals with corporate opportunity: directors and senior executives "shall not, without the consent of a shareholders' meeting or assembly, take advantage of their positions in the company to exploit business opportunities that belong to the company for the benefit of themselves or others, or run the same kind of business as that of the company for themselves or others."

The other key amendment with implications for VC investment is a set of rules that clear the way for the introduction of preferential shares. Those rules are in the following four articles: (1) Article 35: shareholders claim dividends in accordance with their shares of capital, unless they all agree otherwise; (2) Article 43: voting rights in a shareholders' meeting are exercised in accordance with their shares of capital, unless the company by-laws state otherwise; (3) Article 127: each share of the same class is entitled to the same rights; (4) Article 132: the State Council may formulate regulations governing the issuance of classes of shares other than those specified in this law.

Source: Authors.

among other things, what might be allowable and how registration authorities are to accommodate these new legal tools.

Widening the Exit for Venture Investments

Chinese companies have been listed on overseas stock exchanges since the early 1990s. Through the late 1990s, it was essentially only the largest SOEs that listed overseas (almost entirely on the New York Stock Exchange or Hong Kong Stock Exchange). There were few non-state-owned companies that were qualified to list in that period, and for those that were, the government sought to control their ability to list by requiring that they get permission to do so from the securities regulator. However, as the number and quality of private firms increased, many of them with capital from foreign private equity and VC funds, the government also liberalized its position on listing abroad and eventually removed itself entirely from the approval process. That was the situation through the beginning of 2005. The first part of this decade became a bit of a golden age for foreign VC in China because the VC firms had found a model by which they were able to legally invest in Chinese companies and list the successful investments overseas. All of the well-known VC-invested Chinese companies of recent times were divested under that model, and foreign VC funds during that period almost universally targeted the listing of their Chinese portfolio companies on overseas exchanges.

The large changes in the domestic stock markets over the past couple of years have created a fundamentally new situation—for the first time in modern China, private entrepreneurs have seen not just a reason to list their company domestically but also a more realistic prospect that they might actually be able to achieve it. The full implications of this novel situation for VC financing in China cannot yet be known—but they are certainly likely to be positive. An important element of the improved domestic capital market environment has been the creation of the Shenzhen SME Board, which saw 10 VC-backed IPOs in 2006 (although that was only 23 percent of all VC-backed IPOs in 2006; the rest took place in foreign markets). Despite the existence of a stronger IPO avenue for domestic exits, there has been much less progress for one of the other primary channels for exit—strategic sale (that is, merger or acquisition). The revised Company Law still requires that the shares held at the time of the IPO be subject to a one-year lockup. No existing shareholders are allowed to sell at the IPO. That rule significantly increases the minimum holding period for a VC investor.

Acquisitions of Chinese businesses by foreign investors and the establishment of related onshore acquisition vehicles are subject to a multistep, multiagency government approval process, and these approvals depend on the ownership of the target (state-owned, private, or publicly listed), type of transaction, amount invested, and industry involved (Eich and Li 2007). For instance, special approval is required from the central SASAC if one is buying an SOE and the transaction goes beyond the scope of the normal approval process for any acquisition. Because of the various laws that govern foreign investment in China and the required government review and approval, foreign VC firms have chosen to invest in foreign-domiciled holding companies and sell these vehicles to a third party as a completely offshore transaction that would not require any approvals by the Chinese government.

Recommendations

The preferred method of realizing VC investments in China is through an IPO. The domestic markets have grown and matured over the past few years and have increasingly become a desirable exit route for realizing VC investments. However, improvements can still be made to market-based divestments by VC funds, and those improvements are largely within the scope of the CSRC.

First, the time required by the application process could be reduced, and the overall transparency of the domestic listing process could be improved. Reducing government involvement in, and management of, listing volumes would also be an important contribution to efficiency. Indeed, listings have been stopped intermittently and for long periods. The potential for such bans or slow downs to reappear creates considerable uncertainty about the reliability of the domestic stock market as an exit route for VC investment.

Second, the government could consider the further shortening of lock-up periods to the international practice of six months and allowing early investors to sell shares at the IPO stage. VC investors will have invested in a company for several years before its listing and need the maximum flexibility possible to liquidate their positions.

Finally, the government can provide a mechanism for foreign-VC-invested companies to list on foreign as well as local exchanges, a move that would undoubtedly lead to more companies listing in China. In the short term, it is foreign-financed firms that will generally have better practices in terms of equity structuring, governance, accounting transparency, and so on. Allowing such firms to list is the easiest way to produce quick models and also encourage VC firms to look harder at domiciling their funds domestically and investing directly in domestically domiciled corporations.

As for the primary method of divestment in the most mature VC markets—M&As—there are many possible actions to be taken to advance that method in China. The government could move from a system requiring government approval of changes to the shareholder structure (that is, ownership) toward one of registration. Consideration could be given to clarifying what constitutes a strategic industry (in which transactions require a high level of scrutiny and a rigorous system of government approvals) so as to allow the increasing liberalization of the rules for all other transactions. The benefit of having strategic shareholders is well documented and proven worldwide, particularly in innovative industries with intensive knowledge-transfer needs. The current system of government approvals is, to a certain degree, a holdover from the period during which all companies were state-owned. However, as China's economy is increasingly privately owned and market oriented, it needs regulations that allow companies to be more flexible and nimble. The behavior of state-owned companies should be controlled by their shareholders, and the responsibility of all companies and persons to behave legally should be overseen by the rule of law.

The Role of the Government in Supporting the VC Industry

On the one hand, China's VC industry could benefit from government actions to strengthen the VC ecosystem; on the other hand, the government seems to have concentrated on heavy and direct investment in the VC industry, as was discussed earlier in this chapter. This contrast raises the critical issue of the proper role of the government in supporting the VC industry. A thorough investigation of this issue would go beyond the scope of this study, given the scarcity of relevant data in the public domain. Nonetheless, three observations can be made on the basis of existing knowledge.

First, there are some theoretical justifications for government intervention in supporting the VC industry. They begin with the idea that VC supports innovative industries that produce positive externalities that ripple through the economy (Secrieru and Vigneault 2004). Thus, to follow this rationale, government intervention is needed and is appropriate to get nascent businesses off the ground. Government intervention could also help in the process by signaling that the quality of the start-up and early-growth firms is sufficient to warrant other sources of external funding. Finally, government activities in the fiscal, regulatory, and R&D areas can help create the enabling environment for both the VC industry and the potential VC investee companies.

Second, countries have mixed experiences with direct government involvement in the VC industry, and there are some lessons to learn. Many governments around the world have been deeply involved in the development of the VC industry at the early stages and beyond. The efficacy of direct government engagement in the sector is subject to debate; the transparency of government interventions and the academic study of their effects have both been very limited.[62] A few government programs have been successful in promoting the VC industry, while others have resulted in a substantial waste of public resources and have put the VC industry on a worse footing than before. For example, research on one U.S. government program suggests that it did not meet various targets, such as increased geographic distribution of VC funding, and the performance of the investee companies was no different than that by those receiving purely private VC investments (Lerner 1999). Other studies have suggested that government intervention can generate problems in terms of creating market distortions, crowding out the private sector, failing to meet public policy goals, and deploying public resources inefficiently (Lerner 2002). On balance, the evidence seems to imply that the most successful direct government interventions have several features in common. They (1) substantially leverage government funding by requiring as much or more in private funding, (2) use private sector management of the fund, (3) set forth a clear investment objective (for example, a sector development target) with clear return benchmarks, (4) establish a finite period for exiting the investments, (5) are closely coordinated with broader economic development activities as well as other fiscal incentives, such as tax breaks for specific industries, and (6) are geared to improving the general investment climate for the VC industry (including the regulatory framework, tax efficiency, sophisticated infrastructure and financial institutions, and human resource policy).

Third, for China, priority should be given to strengthening the VC ecosystem, and the merits of direct government involvement in the VC industry should be assessed in a broad framework. As with many other countries, the transparency as well as the rigorous empirical study of government direct interventions in China's VC industry have been limited. And such a study would necessarily be complex. There is also a need to go beyond the VC industry to assess the merits of the various modalities of government financing of the early-stage development phases of innovation, from R&D to pilot production. After all, VC financing, as important as it is, has only a limited role to play (as indicated at the outset of this chapter) and is oriented to the commercialization of innovation, not the initial creation of innovation. With a comprehensive assessment, a strategic plan could point the way forward by clearly defining the role of government in the VC industry and in the financing of innovation more broadly.

Conclusions

Within the broad topic of supporting innovation, financing commercialization stands out as a special issue. It requires not only capital, but also external risk capital, that is, capital from outside the firm that is supplied by investors willing and able to take the risks involved in technology creation, adaptation, and adoption. The VC industry emerged to fill the funding gap for start-up and early-stage risk capital for innovative firms.

Despite its relatively early start in the mid-1980s and strong government backing, China's domestic VC industry remains in an early stage of development. Creating a viable VC industry involves more than setting up and capitalizing a number of individual VC funds. More important is the creation of the ecosystem for the industry, the key elements of which include the structure, funding, management, and exit routes for VC investments. The VC industry in China faces challenges in those areas. The way forward is for the government to invest more in improving the ecosystem, as follows:

- *Structure:* With close involvement of institutional investors, assess the new VC structures by conducting an assessment of the operations of those domestic VC funds created following the newly amended Partnership Law and by identifying loopholes and weaknesses that require further legislative or policy actions.
- *Funding:* Expand the sources of VC funding by considering policy measures to allow institutional investors to begin investing in domestic VC institutions. Because the risks of VC investing are high, the first step could be to develop a short- and medium-term action plan that would provide institutional investors with a roadmap for investing in private equity and VC funds.
- *Management:* Build stronger venture partners for investee companies by enhancing corporate governance. The government could organize the formulation of a Code of Conduct for Corporate Governance to facilitate the enforcement of the amended Company Law. It is particularly advisable for the State Council to adopt a regulation to govern the issuance of preference shares.
- *Exit:* Further widen the exit routes for venture investments by providing mechanisms for foreign-VC-invested companies to list on both foreign and local exchanges and continuously improve the domestic listing process. These steps may include a further reduction of application time, greater transparency, and reducing government management of listing volumes.

Finally, a cross-cutting area of the VC industry is the direct role of the government. As is the case in many other countries, there is very limited transparency regarding direct government intervention in China's VC industry and little rigorous empirical study of its effects. Nonetheless, the practical experiences of other countries are mixed. China could consider giving priority to strengthening the VC ecosystem as detailed previously while it simultaneously assesses the merits of direct involvement in the VC industry and considers the most appropriate role for government intervention in innovation financing more generally.

CHAPTER 5

Moving Forward with Actions

China has a remarkable record of achievement in industrialization and development. Over the past three decades, China maintained GDP growth of about 9 percent per year and lifted over 400 million people out of poverty. Entering the 21st century, China is determined to ensure the sustainability of its economic and social development, to which the innovativeness of business enterprises is critical. In 2006, the government of China laid out a strategy of enterprise-led indigenous innovation. A successful implementation of that strategy is crucial to turning China into an "innovative nation." What actions must the government take to achieve that implementation?

The previous chapters have made recommendations around four themes: pursuing a balanced strategy, creating the right incentives, building the capacity of private enterprises, and strengthening the ecosystem for the VC industry. To set those recommendations into an agenda better suited for action, this chapter reorganizes them into three groups that progress from mindset change to broad policy to some special programs for the government to initiate. The chapter concludes with recommendations for further study needed to deepen the understanding of some key issues touched on in the preceding chapters.

Balanced Strategic Thinking

Strategic thinking is important because it guides policies and actions. The strategy of enterprise-led innovation could benefit from changes in the mindset of policy makers at various levels of the government to strike a better balance on a number of key issues. The following viewpoints could be promoted and spread in policy making and public discussions.

1. *The sustainability of development should be taken as the objective of innovation.* Innovation is only a means, not the end. Efforts in promoting innovation should be oriented to serve the higher objective of sustaining China's economic development.
2. *Adaptation and adoption are also innovation.* A proper balance between technological creation and adaptation and adoption must be carefully maintained. As a developing country, China stands to gain a lot from efforts to adapt and adopt technology, efforts that can easily be sidelined in practice.
3. *Let enterprise and the market play their full role before the government intervenes.* Technological innovation is first of all a business matter to be dealt with by enterprises and the market. The government could promote innovation by refraining from involvement in microeconomic decisions on innovation, such as what to innovate and how to innovate. There is certainly a role for the government to play, but it should start at the point at which enterprises and the market cannot do more or better.
4. *There is a need for a stronger focus on the effectiveness and efficiency of R&D spending.* R&D spending alone does not "buy" innovation. Rising R&D investment brings about results in innovation only when it is matched with the needed S&T workforce and other R&D infrastructure. A further increase of China's R&D spending to 2 percent of GDP over the longer term would be desirable, but a stronger focus on the effectiveness and efficiency of R&D spending, especially public R&D spending, is highly advisable.

Innovation-Supporting Policies

The government could promote enterprise-led innovation by enhancing some existing policies, reviewing and changing some others, and initiating a few new ones:

1. *Continuously support the rising of the private sector.* In China's existing national innovation system, state-owned LMEs and RDIs are the main performers of innovation activities. In the future, however, China's success in technological catching-up is likely to rely more on the capacity of its private sector, especially large private firms. Continuous government support for the growth of the young private sector is therefore of strategic importance. The scope of state ownership in business sectors can be further scaled down through means such as the collection of dividends and secondary share offerings.

2. *Continuously enhance corporate governance of large SOEs.* Board governance should be given priority. As board governance is instituted, it is advisable for the government—in particular, SASACs—to delegate more decision-making power to boards of directors. The EVA-based performance evaluation system, to be introduced before 2010, needs to be carefully designed to adequately account for the potential of innovation activities to create long-term value, and take into full consideration that innovative activities are inevitably subject to some rate of failure.

3. *Carry forward reforms to eliminate distortions in prices.* These changes include reform of energy and natural resources pricing and the enforcement of laws and regulations that protect labor rights, consumers' rights, and the environment.

4. *Formulate a special regulation to enforce article 7 of the Anti-Monopoly Law.* That law requires the state to regulate SOE operations to "protect consumers' interest and promote technological progress."

5. *Review the impact of industrial policies on competition.* Industrial policies that hurt competition—for example, by raising barriers to entry and exit—are going to hurt innovation. Anecdotal evidence suggests the possibly extensive existence of such negative effects.

6. *Orient fiscal incentives to encourage pooled R&D efforts.* The formation of research consortia might be one approach to favor. Joint programs with local or foreign HEIs might be another. Thus, tax incentives could be made particularly generous for joint research programs with foreign companies if the incentives are based on the scale of the foreign involvement and the industrial sector that is the focus of the research. That approach would encourage multinational companies, which already benefit from incentives to localize research activities, to work more closely with Chinese firms.

7. *Review the ceiling on tax-deductible training expenditures of enterprises.* The current ceiling of 2.5 percent of the wage bill may

be too low to encourage spending on employee training, especially in view of the high turnover of employees in private SMEs in labor-intensive sectors.

8. ***Allow Chinese institutional investors to start investing in VC funds.*** The government could consider policy measures to allow for institutional investors to begin investing in domestic VC institutions. Recognizing that the risks of VC investing are high, the first step could be to develop a short- and medium-term action plan that would provide a roadmap for institutional investors to following in investing in private equity and VC funds.

9. ***Organize the formulation of a Code of Conduct for Corporate Governance.*** Such a code would cover both state-owned and private companies. In addition to general guidelines, the code should seek to operationalize relevant rules of the 2005 Company Law, such as the concepts of duty of loyalty and duty of care.

10. ***Promulgate a regulation to govern the issuance of preference shares.*** The regulation could spell out, among other things, what might be allowable and how registration authorities are to accommodate these new legal tools.

SME-Specific Programs

China's emerging private sector is populated with SMEs that are run by relatively inexperienced owners and managers and equipped with relatively simple technologies. A range of public interventions has the potential to significantly improve and enlarge the capabilities of private SMEs in technology adoption, adaptation, and creation. Among these interventions are special programs initiated by central and local governments and carefully tailored to local conditions, such as the following:

1. ***Enforcement of the Labor Contract Law.*** Enforcement can be accomplished through, for example, a program of government-sponsored training and knowledge sharing and HR, legal, and other technical advisory services targeting SMEs in a region. The focus can be on the modernization of HR management, protection of labor rights, and the full use of confidentiality agreements and competition restrictions to protect intellectual property and business secrets.

2. ***SME skill development.*** A pilot program could be designed to test the idea of SME skill development centers. Such centers could have a three-part mission: (1) provide SMEs with management and technical

training related to innovation; (2) provide—through close relationships with schools, training institutions, and the labor market—information on the market for various skills and on the premiums demanded for different job categories; and (3) collect and disseminate success stories, especially those from inside China, that describe the management of a skilled labor force and that generally promote an innovation culture.

3. *Innovation brokerages.* Testing the usefulness of innovation brokerage in the Chinese context could provide valuable lessons. The brokerage could be based on the Norwegian TEFT, in which a technological attaché plays an active role in bringing SMEs together with RDIs and other knowledge pools.

4. *Innovation vouchers.* Through pilot projects, innovation vouchers could be distributed by the government to SMEs to cover part of their cost of participating in public-funded R&D activities. They could help increase the interaction between SMEs and RDIs, HEIs, and large companies. In the Netherlands, where such vouchers were pioneered, studies found that the program stimulated SMEs to engage in many additional assignments with public knowledge institutions.

5. *Personnel mobility schemes.* Such schemes facilitate sharing of personnel between private SMEs and knowledge institutions. They provide government support to (1) encourage enterprises to give internships to graduate engineering students or research scientists, (2) allow researchers at HEIs and RDIs to spend time (typically several months) on sabbatical working in enterprises, and (3) allow engineers and technical personnel at enterprises to spend time at HEIs and RDIs. Again, the merits of this instrument can be tested in pilot programs, which could be based on the U.K. experience with its Business Fellowship program.

6. *Industrial association reform.* Industrial associations have a significant role to play in helping SMEs build their innovation capacity. However, China needs to transform industry associations into truly nongovernment organizations. Certain policy incentives, such as tax benefits and competitive grants for not-for-profit organizations, should apply to industry associations. Local or sectoral pilots could be organized to deepen the reform in this area.

7. *Innovation services that are public goods.* Some innovation services are public goods in nature and must be financed by the government. Most of them are essential to technological progress in SMEs. Further strengthening government financial support in this area is desirable. Strengthening could be done at first on a pilot basis through the

establishment of a special fund at the local level, which could be designed specifically to finance SME-related public-goods innovation services. Among the services covered could be information services such as the Shanghai R&D Public Service Platform; services of local technology-diffusion agencies; and the development and dissemination of common technologies and those technologies (and designs, technical solutions, and so on) that are hard to protect from being imitated by others (and thus have a high positive externality).

8. *Enhancement of the MSTQ system.* The measurements, standards, testing, and quality (MSTQ) system provides a special class of services to firms, especially SMEs. Although the system is not necessarily of a public goods nature, its effect on the innovation capacity of SMEs justifies government support. The focus can be on upgrading the physical infrastructure (building new laboratories and purchasing up-to-date equipment) as well as on modernizing MSTQ management structures, business processes, and marketing capabilities.

Key Issues for Further Study

This report has touched upon a number of issues that call for further study because they are critical to a better understanding of the challenges and viable policy options facing China in its endeavor to develop into an "innovative nation." Here is a selection of those key topics:

1. *Evaluation methodology for public R&D spending.* China needs to keep a close watch on the effectiveness and efficiency of its quickly rising R&D spending, especially public spending. Given the complex nature of the output of R&D activities, such monitoring will require a specially developed methodology that is both analytically sound and practical. The core issue for research would be the measurement of innovation performance.

2. *The route to job-creating innovation.* Full employment is as important to China as innovation. On the surface, technological innovation is often perceived as labor saving. However, the real interaction between innovation and job creation is much more complex. For example, the advance of information and communication technology (ICT) in the developed world helps create jobs in the ICT-enabled service sector for low-skilled labor in developing countries. For China, understanding of this issue needs to be deepened by further study of the actual interaction of technological progress and job creation in

the Chinese economy and the exploration of the potential routes to job-creating innovation.

3. *Innovation management of firms.* China's corporate sector already has quite a number of successful innovators. In one sense, China's problem can be seen as making that number greater. Toward that end, studies that examine the experiences of successful Chinese innovators could be useful. For example, what kind of innovation patterns do they follow? How do they allocate their R&D funding? How do they recruit and train R&D managers? What kind of incentive structures work best? Systematic efforts can then be organized to scale up their success.

4. *Development of innovation services.* Evidence suggests that the benefit Chinese SMEs derive from surrounding innovation services remains far from satisfactory. This area deserves in-depth study, which must be sector-specific because the technical and economic nature of innovation services varies greatly. Such a study may lead to a roadmap to help the government speed up the development of this sector.

5. *The role of government in supplying external risk capital.* The activities of the government of China in financing innovation are extensive and widespread, but the degree of transparency regarding the effectiveness and efficiency of such activities remains low. In most of these activities, the government acts as a supplier of external risk capital to firms through, for example, public VC funds. The government needs a specific set of prioritized actions to deepen its understanding of such financing and to improve the policy framework. Thus, in the near term, the government could consider conducting a comprehensive assessment of the role of the government in the VC industry and the most appropriate role for the government in supplying external risk capital more generally. That assessment could then be used to develop a strategy and implementation plan for future financing.

Notes

1. This follows the definition of Aubert (2005).
2. Official Chinese statistics classify industrial enterprises into three size categories: large, medium, and small. Large enterprises are defined as those that meet the following three criteria: (1) 2,000 or more employees, (2) Y 300 million or more in annual sales revenue, and (3) Y 400 million or more in assets. Medium-sized enterprises are those that meet the following three criteria but not those for large enterprises: (1) 300 or more employees, (2) Y 30 million or more in annual sales revenue, and (3) Y 40 million or more in assets. All other enterprises are small.
3. For the period 1978–2002, see CASS (2004}, p. 57. For the period 2003–05, see NBS and MOST (2004, 2005, 2006), table 6-29.
4. China grants three kinds of patents: invention, which is more in the nature of original innovation; utility models; and design.
5. http://www.wipo.int/pct/en/activity/pct_2007.html#P245_14317.
6. A brief overview of the evolution of China's national innovation system is in Liu and Lundin (2007).
7. NBS and MOST (2007), tables 1-5 and 1-9. In 2006, there were 32,647 LMEs and 3,803 RDIs in China. All R&D employment figures are given in terms of full-time equivalents.
8. For the 376 RDIs transformed into enterprises in 1999 and 2000, the "S&T development expenditure" was Y 3.14 billion in 2001 and Y 3.31 billion

in 2002, accounting, respectively, for 1.4 percent and 1.2 percent of the national total (data from a background note prepared for this study by Xinxin Kong in 2008). Assuming that the share of R&D in S&T development expenditures in the 376 transformed RDIs is the same as that of national totals, it then follows that in 2001 and 2002, RDIs contributed perhaps 2.2 percent and 2.0 percent, respectively, of total R&D expenditure of Chinese enterprises.

9. That small role is, however, similar to that in many other economies. For example, the 2006 share of small firms in China's R&D—roughly 18 percent of spending and employment—is about the same as the average for small firms in the OECD countries. In particular, this share was 14 percent in 1990 and 15 percent in 2002 in Japan, and it was 13 percent in 1995 in the Republic of Korea, before rising to 27 percent in 2001 (Lundin and others 2007).

10. According to the definitions in official Chinese statistics, SOE/LMEs comprise state-owned enterprises (SOEs), limited-liability corporations (LLCs), and joint stock corporations (JSCs). The state owns 100 percent of SOEs and more than 50 percent of LLCs and JSCs.

11. http://www.toyota.co.jp/en/ir/library/annual/pdf/2005/index.html; http://www.ford.com/about-ford/investor-relations/company-reports; http://www.pfizer.com/home/; and http://www.microsoft.com/msft/reports/default.mspx.

12. In addition to the central SASAC, which reports to the State Council, there are SASACs at the provincial and municipal levels as well, performing the same function in SOEs in the portfolios of local governments.

13. *People's Daily*, May 15, 2006.

14. Lundin and others (2007) have also found that, for most ownership groups in China, the patent output of small firms is larger than that of LMEs, and that the patent performance of domestic private firms is superior to that of SOEs across all categories of firm size. The superior performance of private LMEs over state LMEs suggests that ownership may have been a more influential factor than firm size.

15. The experiment was formally launched in June 2004 when SASAC issued a notice to all central SOEs. http://www.chinalawedu.com/news/1200/21829/2006/3/pa159319265419360021 4268-0.htm). Seven pilot SOEs were in the first group. http://cs.xinhuanet.com/gz/07/200801/t20080121_1352293_4.htm.

16. For SOE profitability prior to 2007, see MOF (2007). For total pretax profit in 2007, see http://news.xinhuanet.com/english/2008-01-24/content_7485388.htm. For GDP data, see NBS (2007a).

17. See the following Web sites for the World Bank's previous reports on this issue: http://www.worldbank.org.cn/Chinese/content/636m63551169.shtml; and http://www.worldbank.org.cn/Chinese/content/china_05_07.pdf.

18. http://www.acfic.org.cn/cenweb/portal/user/anon/page/AcficCEWorkd DynamicPage.page?appId=00000000000000000125&categoryCode=040.

19. For example, see Chen (2006, pp. 41–55), and Zhang (2006).

20. The Chinese electrolytic aluminum industry is a case in point. Over the past few years, the government has made a great effort to curb the investment boom in this sector, which suffers from overinvestment and excess capacity. Nonetheless, until October 2007, the government subsidized investment in the sector by offering discounted prices for electricity to a handpicked group of producers. Because electricity accounts for 40 percent of the production cost of the industry, the subsidies helped create the imbalances that the government was trying to correct. See http://finance.sina.com.cn/stock/s/20071012/03194052484.shtml.

21. http://www.ndrc.gov.cn/xwzx/xwtt/t20071207_177938.htm.

22. This section draws on Yusuf and Nabeshima (2007).

23. VAT rebates on semiconductors offered after June 2000 and amended in 2001 were ended in April 2005 (Sigurdson 2005, p. 141).

24. The contribution of research consortia in Japan has been described by Branstetter and Sakakibara (1998, 2002). Such consortia have also been created in Korea and in the United States (see Sakakibara and Branstetter 2003; Sakakibara and Cho 2002). Compared with those efforts, the contribution of foreign firms in China has not been as strong, but Whalley and Xin (2006) find that foreign companies and joint ventures that, on balance, were more capital, technology, and skill intensive were responsible for nearly 60 percent of exports and close to 40 percent of China's growth in 2003–04. They were also responsible for close to one-half of all patent applications to China's patent agency in 2005 and for nearly two-thirds of all patents granted (WIPO 2006).

25. http://www.chinaacc.com/new/63/73/130/2006/2/yi90591111195126002 451-0.htm.

26. http://www.gov.cn/zwhd/2007-03/16/content_552435.htm. The seven criteria are (1) being in line with national laws, regulations, and industrial policies; (2) having self-owned IPR; (3) having a self-owned (*zizhu*) brand; (4) being highly innovative; (5) being produced with advanced technology and being among the internationally advanced in comparison with similar products; (6) having reliable quality after being tested by qualified authorities; and (7) having profitable potential and good market demand, or being able to substitute for imports. The four steps are (1) the applicant (any enterprise or public service unit with a legal presence in China) submits an application to a provincial bureau of S&T; (2) the provincial bureau reviews the application, organizes an expert review, and makes a recommendation to MOST; (3) MOST, in collaboration with NDRC, MOF, and other central government agencies, conducts

the review and verification to come up with a draft catalog; and (4) MOST publishes the draft catalog for public consultation and releases the final catalog with NDRC and MOF if no objection is received.

27. http://www.gov.cn/ztzl/kjfzgh/content_883671.htm.

28. http://www.gov.cn/zwgk/2008-01/15/content_858687.htm.

29. http://www.gov.cn/zwgk/2008-01/15/content_858687.htm.

30. In Germany, industry-level standards have also encouraged upgrading. "The Solingen Law established in 1938, for example, set rigid standards for the quality of cutlery and the right to use the Solingen name (the city where the German cutlery industry is concentrated). The law, a response to practices that were causing product quality to deteriorate, has proved to be an important device for preserving German differentiation" (Porter 1990, pp. 647–48).

31. This assessment covers standards for EVD (for digital audio/video format), AVS (advanced visual systems), TD-SCDMA (for 3G mobile phones), RFID (radio frequency identification), and IGRS (networking for home entertainment systems) (Suttmeier, Yao, and Tan 2007, cited in Zhang 2007b).

32. http://www.zgbfw.com/info/pump-news-189101.html.

33. In Japan, "in sewing machines, for example, standards were established for parts in the early post–World War II period. This spawned numerous parts suppliers, lowered entry barriers into sewing machine assembly, and speeded up attention to new features and quality. In television sets, facsimile, and other products, relatively rapid agreement on standards benefited Japanese industry in progressing to the rapid introduction of new models and features" (Porter 1990, p. 653).

34. These enterprises consist of sole proprietorships and partnerships; and limited liability companies and joint stock companies with private ownership of more than 50 percent.

35. The NBS collects and releases regular data on all industrial enterprises owned and controlled by the state and on those non-state-owned industrial enterprises that have annual sales revenue of more than Y 5 million, an amount known as the "cut-off scale." It is useful to gain a sense of the size of those that are left out by this cut-off. In 2004, the latest year for which data are available, 902,647 private industrial enterprises were in operation. Of that group, the firms above the cut-off scale numbered 119,357, or 13 percent; they accounted for 47 percent of employment and 71 percent of gross value of industrial output. Those below the cut-off scale thus represented 87 percent of the firms and 53 percent of employment but only 29 percent of gross value of industrial output (NBS 2006, pp. 505, 538). Output has more than one measuring indicator, specifying the indicator in a reader-friendly manner to other researchers.

36. The figure also suggests that the larger the share of private firms in a sector, the larger their average size tends to be relative to other firms in the same sector.

The exception is sector 39, water supply, in which private firms appear to be relatively large despite a tiny market share.

37. Data are from a survey conducted in 2005 by the All China Federation of Industry and Commerce (ACFIC) and the State Administration of Industry and Commerce (SAIC) in two tracks. ACFIC randomly drew 2,360 firms from a stratified sample of 429,090 firms in 31 provinces, while SAIC distributed questionnaires to 1,600 firms in areas of 15 provinces where it conducts regular monitoring (ACFIC 2007, p. 210).

38. Indeed, in a reasonably well-functioning market for labor, that kind of shortage would be the point at which such firms should stop hiring more workers.

39. *Enforcement Regulation of the Corporate Tax Law of the People's Republic of China*, enacted by the State Council on December 2007, article 42. http://www.gov.cn/zwgk/2007-12/11/content_830645.htm.

40. http://strategis.ic.gc.ca/engdoc/main.html.

41. http://www.sgst.cn/index.htm.

42. Author's interview with the management of the Platform in May 2008.

43. A service is a public good when (1) consumption of it by one consumer does not affect the consumption of it by another consumer, and (2) it is practically impossible to exclude a consumer who does not pay for the service from consuming it.

44. MOST (2006, 2008) further reports that by 2006, there were 1,331 productivity promotion centers among all service providers and 548 incubators (excluding 62 university high-tech parks). But data for total innovation service providers and employees in 2006 are not available.

45. One of the most famous examples is the funding of Christopher Columbus's 1492 voyage and exploration, funded mostly by King Ferdinand and Queen Isabella of Spain in return for the treasures, lands, and revenues gained through the voyage.

46. In the first stage, seed capital is used to fund initial product research and development and to assess the commercial potential of ideas; similarly, start-up capital helps companies begin the production, marketing, and sale of their products. Early-growth capital helps the company expand its manufacturing and distribution capacity and also fund further R&D. Expansion funding assists a firm in expanding the sale of the product. Later-stage funding can be for buying and selling operations or further expansion (PricewaterhouseCoopers 2008).

47. Definitions are from the glossary on the Web site of the European Private Equity and Venture Capital Association. http://www.evca.eu/toolbox/glossary.aspx?id=982.

48. European Private Equity and Venture Capital Association, "2007 European Private Equity Activity Survey." http://www.evca.eu/uploadedFiles/Home/

Knowledge_Center/EVCA_Research/Latest_Data/Activity_Slides_Prelimina ry2007.ppt. Note that VC investments were only 8.4 percent of all private equity investments in Europe in 2007. Also, the largest non-EU investor was the United States, representing 26.4 percent of all new investment in private equity and VC in Europe in 2007.

49. MII news release, February 7, 2007.

50. For information on the Yozma Group, visit http://www.yozma.com/ overview/.

51. Hogan and Hartson, "China Update," July 2007. http://www.hhlaw.com/files/ Publication/40e685cc-28b8-4207-97d5-024fa8b2d82f/Presentation/ PublicationAttachment/0b93210b-18ac-4305-adbe-375c3ab03bde/China Update_July2007.pdf.

52. One legacy of the SOE reforms was that a large amount of shares on Chinese domestic stock markets were held by the government or government-owned enterprises and were not tradable—more than 60 percent of A-shares were not tradable in 2004. The China Securities Regulatory Commission engineered a "3 for 10" solution, in which holders of nontradable shares agreed to give 3 shares out of their own holdings at no charge to holders of tradable shares for every 10 tradable shares they held. The nontradable share reform process was completed in 2006.

53. Shenzhen Stock Exchange. http://www.szse.cn/main/en/SMEBoard/ smeboarddata/.

54. Zero2IPO. http://www.zero2ipo.com.hk/china_this_week/detail.asp?id=4516.

55. The first and second VC firms in the form of limited partnership were reportedly created in Shenzhen. http://dycj.ynet.com/article.jsp?oid=25292037. In Wenzhou, Zhejiang province, a private equity firm in the form of a limited partnership, Donghai Venture, was established in August 2007. Its general partner is led by a former official of the China Securities Regulatory Commission, and all of Donghai's LPs are private businesses in Wenzhou. http://finance.sina.com.cn/chanjing/b/20070827/02173917478.shtml.

56. Speech by Gao Xiqing, Vice Chairman of the NSSF, at the JP Morgan China Conference, Grand Hyatt Hotel, Beijing, April 26, 2007.

57. *China Business News*, February 5, 2007.

58. China Insurance Regulatory Commission. http://www.circ.gov.cn.

59. The NSSF was directed by the State Council, to which it reports, to make two exceptions, the most important of which was a large investment in the Bohai Industrial Fund—a new fund established to back projects in the Tianjin area. The Bohai fund has heavy political backing, and the other investors are all large state-owned enterprises and institutions. It has not started making investments yet. The other exception was an investment in the Sino-Belgian Direct Equity Investment Fund, also a politically motivated fund.

60. The limits of investment in alternatives (including VC) could consist of some maximum percentage of (1) an institution's total assets, (2) paid-up capital and reserves, or (3) the shareholding of the investee fund or some combination thereof. The regulatory guidelines should specify the exposure limits as well as the valuation, classification, risk weightings, approval processes, and so on, for investments in this asset class by regulated financial institutions. For illustrations of such guidelines, the Reserve Bank has "Prudential Guidelines on Bank's Investment in Venture Capital Funds." http://rbidocs.rbi.org.in/rdocs/content/PDFs/72128.pdf; and the Monetary Authority of Singapore has its "Notice to Banks on Private Equity and Venture Capital Investments." http://www.mas.gov.sg/legislation_guidelines/banks/notices/Notice_630__Private_Equity_And_Venture_Capital_Investments.html.

61. http://www.csrc.gov.cn/n575458/n870399/n1337892/n3883477/3892788.html.

62. For example, a November 2007 report on public intervention in the Australian VC market has no real conclusion on effectiveness (Lerner and Watson 2007).

References

Aghion, Philippe. 2006. "A Primer on Innovation and Growth." http://www.bruegel.org/Files/media/PDF/Home/Primer_Innovation_Growth_13Oct.pdf.

ACFIC (All China Federation of Industry and Commerce). 2007. *The Large Scale Surveys on Private Enterprises in China*. Beijing: China Industry and Commerce Association Press.

Amiti, Mary, and Caroline Freund. 2008. "The Anatomy of China's Export Growth." Policy Research Working Paper 4628, World Bank, Washington, DC.

Asheim, Bjorn Terje, Arne Isaksen, Claire Nauwelaers, and Franz Todtling. 2003. *Regional Innovation Policy for Small-Medium Enterprises*. Cheltenham, U.K.: Edward Elgar.

Aubert, Jean-Eric. 2005. "Promoting Innovation in Developing Countries: A Conceptual Framework." Working Paper 3554, World Bank, Washington, DC.

Auerswald, Philip E., and L. M. Branscomb. 2003. "Valleys of Death and Darwinian Seas: Financing the Invention to Innovation Transition in the United States." *Journal of Technology Transfer*, 28 (3–4): 227–39.

Baumol, William. 2002. *The Free-Market Innovation Machine*. Princeton, NJ: Princeton University Press.

Baumol, William J., Robert E. Litan, and Carl J. Schramm. 2007. *Good Capitalism, Bad Capitalism, and the Economics of Growth and Prosperity*. New Haven, CT: Yale University Press.

Bottelier, Pieter. 2004. "Venture Capital and Innovation in China." Lecture series, Johns Hopkins University, School for Advanced International Studies, Washington, DC.

Branstetter, Lee, and Mariko Sakakibara. 1998. "Japanese Research Consortia: A Microeconometric Analysis of Industrial Policy." *Journal of Industrial Economics* 46 (2): 207–33.

———. 2002. "When Do Research Consortia Work Well and Why? Evidence from Japanese Panel Data." *American Economic Review* 92 (1): 143–59.

CASS (Chinese Academy of Social Sciences/Institute of Industrial Economics). 2004. *China's Industrial Development Report*. Beijing: Economy and Management Publishing House.

Center for Private Equity and Entrepreneurship. 2005. "Entrepreneurial Exit Strategies Program: Results of Survey of Private Equity Funds." Tuck School of Business at Dartmouth, Hanover, NH. http://mba.tuck.dartmouth.edu/pecenter/research/pdfs/exits_survey.pdf.

Centre for Asia Private Equity Research. 2007. "Asia Private Equity Review 2007 Year End Review." http://www.asiape.com/?Publications:Asia_Private_Equity_Review:APER0712YE.

Chen, Qingtai. 2006. *Some Issues of Enterprises' Indigenous Innovation. Comparative Studies, vol. 28*. Beijing: CITIC Press.

CPB (Netherlands Bureau for Economic Policy Analysis). 2007. "The Effects of the Dutch Innovation Voucher 2004 and 2005." CPB Document 140, The Hague. http://www.cpb.nl/eng/pub/cpbreeksen/document/140/doc140_summary.pdf.

CSRC (China Securities Regulatory Commission). 2002. "Guiding Principles of Corporate Governance of Listed Companies." Beijing. http://www.csrc.gov.cn/n575458/n870399/n1337892/n3883477/3892788.html.

———. 2008. "China Capital Market Development Report." Beijing. http://www.csrc.gov.cn/n575458/n4001948/n4002030/n9434750.files/n9434747.pdf.

Dahlman, Carl, Douglas Zhihua Zeng, and Shuilin Wang. 2007. *Enhancing China's Competitiveness through Lifelong Learning*. Washington, DC: World Bank.

Deloitte Research. 2006. "Seven Disciplines for Venturing in China." Deloitte, New York. http://www.deloitte.com/dtt/cda/doc/content/US_SevenDisciplines China_research.pdf.

DIE/DRC (Department of Industrial Economics of the Development Research Center of the State Council) and others. 2008. *Annual Report on Automotive Industry in China (2008)*. Beijing: Social Sciences Acamedic Press.

Dodgson, Mark, John Mathews, and Tim Kastelle. 2006. "The Evolving Role of Research Consortia in East Asia." *Innovation: Management, Policy and Practice*. 8 (1–2): 84–101.

Dotzler, Fred. 2001. "What Do Venture Capitalists Really Do, and Where Do They Learn to Do It?" *Journal of Private Equity*. Winter. http://www.denovovc.com/arti cles/2001_Dotzler.pdf.

Dougherty, Sean, Richard Herd, and Ping He. 2007. "Has a Private Sector Emerged in China's Industry? Evidence from a Quarter of a Million Chinese Firms." *China Economic Review* 18 (07): 309–34.

Drucker, Peter. 1985. *Innovation and Entrepreneurship: Practice and Principles*. New York: Harper and Row.

Dutz, Mark. 2007. *Unleashing India's Innovation Potential*. Washington, DC: World Bank.

EC (the European Commission). 2002. "Directory of Measures in Favor of Entrepreneurship and Competitiveness." Enterprise and Industry Department. http://ec.europa.eu/enterprise/enterprise_policy/charter_directory/en/tech nology/norway.htm.

Eich, Daniel Patrick, and Chuan Li. 2007. "Private Equity Investments in China: Impact of Recent Legal Reforms." *Venture Capital Review* 18 (Winter). National Venture Capital Association, Arlington, VA; Ernst and Young, Washington; DC. http://www.kirkland.com/siteFiles/kirkexp/publications/2261/Document1/Recent_Legal_Reforms.pdf.

Ernst & Young. 2005. "Creating a Technology Hotbed in China: Lessons Learned from Silicon Valley and Israel: Summary Report and Suggestions to the CVCA." September. http://www.ey.com/Global/Assets.nsf/Austria/Venture_Capital_in_China_0905/$file/Venture_Capital_in_China.pdf.

Ernst and Young Venture Capital Advisory Group. 2006. "2Q 2006 Venture Insights," presented at Milestones Asia 2006 Conference, Beijing, October 24.

Gill, Indermit, and Homi Kharas. 2007. *An East Asian Renaissance: Ideas from Economic Growth*. Washington, DC: World Bank.

Goldberg, Itzhak. 2004. "Poland and the Knowledge Economy—Enhancing Poland's Competitiveness in the European Union." World Bank, Washington, DC, September. http://www-wds.worldbank.org/servlet/main?menuPK=64187510&pagePK=64193027&piPK=64187937&theSitePK=523679&entit yID=000012009_20041007160143.

Gompers, Paul A., and Josh Lerner. 1999. *The Venture Capital Cycle*. Cambridge, MA: MIT Press.

Griliches, Zvi. 1998. *R&D and Productivity*. Chicago: University of Chicago Press.

Hall, Bronwyn H. 2005. "The Financing of Innovation." University of California at Berkeley. http://emlab.berkeley.edu/users/bhhall/papers/ShaneHB_BHH%20chapter_rev.pdf.

Higgins, Andrew. 2004. "As China Surges, It Also Proves a Buttress to American Strength." *Wall Street Journal*, January 30. http://online.wsj.com/article/SB107542341587316028.html.

Hill, Linda A., Tarun Khanna, and Emily A. Stecker. 2007. "HCL Technologies (A)." Harvard Business School, Cambridge, MA. http://doi.contentdirections. com/mr/hbsp.jsp?doi=10.1225/408004.

HKMA (Hong Kong Monetary Authority). 2006. "How Efficient Has Been China's Investment?" http://www.info.gov.hk/hkma/eng/research/RM19-2006.pdf.

Hou, Chi-Ming, and San Gee. 1993. "National Systems Supporting Technical Advance in Industry: The Case of Taiwan." In *National Innovation Systems: A Comparative Analysis*, ed. Richard R. Nelson, 384–413. New York: Oxford University Press.

Huang, Can, and others. 2005. "Organizing, Program, and Structure: An Analysis of the Chinese Innovation Policy Framework." Working Papers de Economia (Economics Working Papers), Number 17. Departamento de Economia, Gestão e Engenharia Industrial, Universidade de Aveiro, Portugal.

IFC (International Finance Corporation). 2006. "China: Anhui Conch Cement: Investment Review Memorandum." IFC, Washington, DC.

Kim, Linsu. 1993. "National System of Industrial Innovation: Dynamics of Capability Building in Korea." In *National Innovation Systems: A Comparative Analysis*, ed. Richard R. Nelson, 357–83. New York: Oxford University Press.

Kuijs, Louis. 2006. "How Will China's Saving-Investment Balance Evolve?" Working Paper 3958, World Bank, Washington, DC.

Lake, Rick, and Ronald A. Lake. 2000. *Private Equity and Venture Capital: A Practical Guide for Investors and Practitioners*. London: Euromoney Books.

Lerner, Josh. 1999. "The Government as Venture Capitalist: The Long-Run Impact of the SBIR Program." *Journal of Business* 2 (3): 285–318.

———. 2000. *Venture Capital and Private Equity*. New York: John Wiley.

———. 2002. "When Bureaucrats Meet Entrepreneurs: The Design of Effective 'Public Venture Capital Programs.'" *Economic Journal* 112 (477): F73–F84. http://papers.ssrn.com/sol3/papers.cfm?abstract_id=308910.

Lerner, Josh, and Brian Watson. 2007. "The Public Venture Capital Challenge: The Australian Case." Harvard Business School and Georgica Associates. http:// papers.ssrn.com/sol3/papers.cfm?abstract_id=1027445.

Li, Wei, and Colin Xu. 2007. "Chinese Firms' Innovation Performance: Existing and New Evidence." Background paper, World Bank, Washington, DC.

Library of Congress Country Studies. 1997. "Brazil: The Computer Industry Policy." http://www.photius.com/countries/brazil/national_security/brazil_national_ security_the_computer_industr~290.html.

Lin, Yifu. 2007. "Indigenous Innovation Should Reflect Comparative Advantage." *Caijing Magazine*, February 5.

Liu, Xielin, and Nannan Lundin. 2007. "Toward a Market-Based Open Innovation System of China." http://www.globelicsacademy.net/2007/papers/Xielin%20 Liu%20Paper%201.pdf.

Liu, Yingqiu, and Zhixiang Xu, ed. 2006. *Report on Competitiveness of Chinese Private Enterprise, No. 3.* Beijing: Social Sciences Academic Press.

———. 2007. *Report on Competiveness of Chinese Private Enterprises, No. 4: Human Capital and Competitiveness Index.* Beijing: Social Sciences Academic Press.

Lundin, Nannan, Fredrik Sjoholm, He Ping, and Jinchang Qian. 2007. "The Role of Small Firms in China's Technology Development." Working Paper 695, Research Institute of Industrial Economics, Stockholm, Sweden.

Luzio, Eduardo. 1996. *The Microcomputer Industry in Brazil: The Case of a Protected High-Technology Industry.* Westport, CT: Praeger.

Mackenzie, Davin. 2007. "China: Financing for Innovation, Recommendations for Improving the Investment Environment for the Domestic Venture Capital Industry." Background paper prepared for the World Bank, Washington, DC.

Martin, Stephen, and John Scott. 1999. "The Nature of Innovation Market Failure and the Design of Public Support for Private Innovation." *Research Policy* 29 (4-5): 437–47.

Mathews, John A., and T.S. Poon. 1995. "Technological Upgrading through Alliance Formation: The Case of Taiwan's New PC Consortium." *Industry of Free China* 74 (6): 43–58.

McGregor, Jena. 2007. "The Employee Is Always Right." *Business Week,* November 19.

MOF (Ministry of Finance). 2007. *China Fiscal Yearbook.* Beijing: China Finance Publishing House.

MOST (Ministry of Science and Technology). Various years. *China Science and Technology Development Report 2005.* Beijing: Science and Technology Literature Press.

MOST Study Team. 2006. *Investigation Report on Indigenous Innovation Capacity of Industries of Our Country.* Beijing: Science Press.

Motohashi, Kazuyuki. 2006. "China's National Innovation System Reform and Growing Science Industry Linkage." *Asian Journal of Technology Innovation* 14 (2) (2006): 49–65.

National Venture Capital Association. 2007. "Money Tree Report: Q4 2007 / Full Year 2007." https://www.pwcmoneytree.com/MTPublic/ns/moneytree/filesource/exhibits/National_MoneyTree_full_year_Q4_2007_Final.pdf.

———. 2008. "News Release on Fourth Quarter Performance of Venture Capital," April 28. http://www.nvca.org/pdf/Q407PerformanceReleaseFINAL.pdf.

NBS (National Bureau of Statistics). 2006. *China Statistical Yearbook 2006.* Beijing: China Statistics Press.

———. 2007a. *China Statistical Yearbook 2007.* Beijing: China Statistics Press.

———. 2007b. *China Labor Statistical Yearbook 2007.* Beijing: China Statistics Press.

NBS and MOST. Various years. *China Statistical Yearbook on Science and Technology*. Beijing: China Statistics Press.

NBS and NDRC (National Development and Reform Commission). Various years. *Statistics on Science and Technology Activities of Industrial Enterprises*. Beijing: China Statistics Press.

NBS, NDRC, and MOST. 2007. *China Statistics Yearbook on High Technology Industry*. Beijing: China Statistics Press.

NDRC. 2006. "Special Development Plan for Cement Industry." http://www. sdpc.gov.cn/gyfz/zcfg/t20061019_89101.htm.

Nelson, Richard R., and Nathan Rosenberg. 1993. "Technical Innovation and National Systems." In *National Innovation Systems: A Comparative Analysis*, ed. Richard R. Nelson, 3–22. New York: Oxford University Press.

Odagiri, Hiroyuki, and Akira Goto. 1993. "The Japanese System of Innovation: Past, Present, and Future." In *National Innovation Systems: A Comparative Analysis*, ed. Richard R. Nelson, 76–114. New York: Oxford University Press.

OECD (Organisation for Economic Co-operation and Development). 2002. "The Size of Procurement Markets." http://www.oecd.org/dataoecd/34/14/ 1845927.pdf.

———. 2004. *Global Knowledge Flows and Economic Development*. Paris: OECD.

———. 2005a. *Corporate Governance of State-Owned Enterprises: A Survey of OECD Countries*. Paris: OECD.

———. 2005b. *OECD SME and Entrepreneurship Outlook*. Paris: OECD.

———. 2007. *Review of Innovation Policy China: Synthesis Report*. Released in Beijing, August 27.

PBOC (People's Bank of China). 2007. "Summary of Sources and Uses of Funds of Financial Institutions." http://www.pbc.gov.cn/english/diaochatongji/tongj ishuju/gofile.asp?file=2007S01.htm.

Porter, Michael E. 1990. *The Competitive Advantage of Nations*. New York: Free Press.

PricewaterhouseCoopers. 2008. "Money Tree Report." https://www.pwcmoneytree. com/MTPublic/ns/nav.jsp?page=definitions#stage.

Roper, Stephen, Jim H. Love, Jonathan Scott, Phil Cooke, Nick Clifton, and Nola Hewitt-Dundas. 2007. "The Scottish Innovation System: Review and Application of Policy." Department of Enterprise, Transport and Lifelong Learning, Scottish Executive, Edinburgh. http://www.scotland.gov.uk/ Publications/2007/03/19165356/0.

Sakakibara, Mariko, and Dong Sung Cho. 2002. "Cooperative R&D in Japan and Korea: A Comparison of Industrial Policy." *Research Policy* 31 (5): 673–92.

Sakakibara, Mariko, and Lee Branstetter. 2003. "Measuring the Impact of U.S. Research Consortia." *Managerial and Decision Economics* 24 (2–3): 51–69.

Schoonmaker, Sara. 2002. *High-Tech Trade Wars: U.S.-Brazilian Conflicts in the Global Economy.* Pittsburgh, PA: University of Pittsburgh Press.

The Scottish Government. 2007. "The Scottish Innovation System: Review and Application of Policy." Scotland, March 26.

Secrieru, Oana, and Marianne Vigneault. 2004. "Public Venture Capital and Entrepreneurship." Working Paper 2004-10, Bank of Canada, Ottawa. http://www.bankofcanada.ca/en/res/wp/2004/wp04-10.pdf.

Serger, Sylvia S., and Magnus Breidne. 2007. "China's Fifteen-Year Plan for Science and Technology: An Assessment." *Asia Policy* 4: 139.

Sigurdson, Jon. 2005 *Technological Superpower China.* Cheltenham, U.K.: Edward Elgar.

Union Square Ventures. 2007. "Failure Rates in Early Stage Venture Deals." Union Square Ventures, New York. http://www.unionsquareventures.com/2007/11/failure_rates_i.html.

Wang, Qingsong. 2007. "China Introduces Tax Incentives for Venture Capital Investments in High-Tech Sector." Orrick, Herrington, and Sutcliffe LLP, Beijing. http://www.orrick.com/fileupload/1164.pdf.

Wang, Yuan, Weizhong Wang, and Gui Liang, ed. 2007. *China Venture Capital Development Report 2007.* Beijing: Economy and Management Publishing House.

Whalley, John, and Xian Xin. 2006. "China's FDI and Non–FDI Economies and the Sustainability of Future High Chinese Growth." Working Paper 12249, National Bureau of Economic Research, Cambridge, MA.

WIPO (World Intellectual Property Organization). 2006. *WIPO Patent Report: Statistics on Worldwide Patent Activity (2006 Edition).* Geneva: WIPO.

———. 2007. "The International Patent System Yearly Review: Development and Performance in 2007." http://www.wipo.int/export/sites/www/pct/en/activity/pct_2007.pdf.

World Bank. 2004. "Chile—New Economy Study." Policy Note, Report 25666, World Bank, Washington, DC.

———. 2005. *Deepening Public Service Unit Reform to Improve Service Delivery.* Report 32341, Poverty Reduction and Economic Management Unit, East Asia and Pacific Region, World Bank, Washington, DC. http://www-wds.worldbank.org/external/default/WDSContentServer/WDSP/IB/2005/08/01/000012009_20050801102726/Rendered/PDF/32341rev01CHA.pdf.

Wu, Ceng, and Changlin Gao. 2007. "On the Innovation of Hi-Tech Industries from a Technology Dependence Perspective." http://www.sts.org.cn/fxyj/zcfx/documents/0706121.htm.

Wu, Jinglian. 2007. "Does China Need to Change Its Industrialization Path?" In *East Asian Visions: Perspectives on Economic Development,* Indermit Gill,

Yukon Huang, and Homi Kharas ed., 285–308. Washington, DC: World Bank.

Yusuf, Shahid, and Kaoru Nabeshima. 2007. "Strengthening China's Technological Capacity." Working Paper 4309, World Bank, Washington, DC.

Yusuf, Shahid, Shuilin Wang, and Kaoru Nabeshima. 2009. "Fiscal Policies for Innovation." In *Innovation for Development and the Role of Government: A Perspective from the East Asia and Pacific Region*, Qimiao Fan, Kouqing Li, Douglas Zhihua Zeng, Yang Dong, and Runzhong Peng, ed., 149–80. Washington, DC: World Bank.

Zero2IPO. 2007. *China Venture Capital Annual Report 2007*. Beijing: Zero2IPO Group. http://www.zero2ipo.com.hk/research/reports.asp?ReportType=1.

Zhang, Wenkui. 2007a. "Indigenous Innovation of the State-Owned Enterprises: Current Situation, Problems, and Suggestions." Background paper, World Bank, Washington, DC.

———. 2007b. "Promoting Demand on Innovation Products through Government Procurement, Standard Formation, and Orientation Measures." Background paper, World Bank, Washington, DC.

Zhang, Yongwei. 2006. "Why Do Enterprises Lack Motivations for Innovation?" http://www[GF2].gzii.gov.cn in Chinese only.

Index

Boxes, figures, notes, and tables are denoted by b, f, n, and t, following the page numbers.